Synchrotron Radiation in Chemistry and Biology III

AKADEMIE-VERLAG BERLIN

Topics in Current Chemistry · Vol. 151

Synchrotron Radiation in Chemistry and Biology III

Synchrotron Radiation in Chemistry and Biology III

With 127 Figures and 3 Tables

AKADEMIE-VERLAG BERLIN
1990

Die Originalausgabe erscheint im Springer-Verlag
Berlin · Heidelberg · New York · London · Paris · Tokyo · Hongkong
in der Schriftenreihe "Topics in Current Chemistry", Volume 151

Vertriebsrechte für die sozialistischen Länder:
Akademie-Verlag Berlin

Vertriebsrechte für alle Staaten mit Ausnahme der sozialistischen Länder:
Springer-Verlag Berlin · Heidelberg · New York · London · Paris · Tokyo · Hongkong

ISBN 3-540-51201-2 Springer-Verlag Berlin · Heidelberg · New York
ISBN 0-387-51201-2 Springer-Verlag New York · Berlin · Heidelberg

ISBN 3-05-500769-7 Akademie-Verlag Berlin

Erschienen im Akademie-Verlag Berlin, Leipziger Straße 3—4, Berlin, DDR-1086
Lizenznummer: 202 · 100/527/90
Printed in the German Democratic Republic
Gesamtherstellung: VEB Druckerei „Thomas Müntzer", 5820 Bad Langensalza
LSV 1275
Bestellnummer: 764 083 0 (3072/151)
14800

Guest Editor

Editorial Board

Preface

This is the last volume of a set of three within "Topics in Current Chemistry" devoted to applications of synchrotron radiation in chemistry and biology (the two previous ones are Vol. 145 and 147). It contains nine contributions dealing with novel developments in X-ray diffraction, X-ray spectroscopy, and other technical developments. In the first chapter K. C. Holmes gives a vivid acount of the "distant" past, describing the early developments in X-ray diffraction from muscles that lead to the establishement of one of the first synchrotron laboratories for biological applications (the EMBL Outstation at DESY, Hamburg). The volume ends with a view into the future by C. Riekel who surveys some experimental possibilities envisaged for the European Synchrotron Radiation Facility (ESRF) to be built in Grenoble. The present state of the art is described in the remaining chapters which deal with biological and chemical applications at synchrotron laboratories in Orsay, Daresbury, Hamburg, and Cornell. Several contributions are concerned with supramolecular assemblies such as biological or synthetic polymers (protein fibers, Ch. 2; DNA, Ch. 3; polyethylene and others, Ch. 6) or liquid crystals and membranes (Ch. 5). Chapter 4 describes the revival of Laue diffraction methods and how it can be applied to protein crystallography. The approach depends on the wavelength spread of synchrotron radiation, and this property also forms the basis for the studies of Ch. 7 and 8 dealing with atomic and molecular interactions. The contributions to this and the previous volumes illustrate the diversity of synchrotron radiation research, even when we consider only the biological and chemical applications. It is therefore not possible to define a unifying topic other than the use of synchrotron light, and this is in fact one of the strengths of this new discipline. Thus each chapter represents an independent unit.

I would like to express my thanks to the authors who agreed to contribute in spite of their busy schedules. I am especially indebted to Elke Spader whose organizational talent and persistence were invaluable in putting the three volumes together. Finally, it was a pleasure to collaborate with Dr. Stumpe and Mrs. Frank from Springer-Verlag.

Hamburg, February 1989 Eckhard Mandelkow

Table of Contents

Table of Contents of Volume 145

Table of Contents of Volume 147

Synchrotron Radiation as a Source for X-Ray Diffraction The Beginning

Kenneth C. Holmes

Max-Planck-Institut für medizinische Forschung, Abt. Biophysik, Jahnstr. 29, 6900 Heidelberg, FRG

The initial motivation for the development of synchrotron radiation as an X-ray source derived from attempts to understand the molecular mechanism of muscle contraction. In the early sixties there was renewed interest in the detailed low angle X-ray diffraction patterns given by muscle fibres. These had first been examined by H. E. Huxley more than a decade before but had been put on one side because of the great success of the electron microscopic methods of ultra thin sectioning and negative staining which yielded such rich rewards [1, 2]. Huxley had shown that the low angle equatorial reflexions were a sensitive function of the physiological state of the muscle. In 1965, Reedy, Holmes, and Tregear showed that not only the equatorial reflexions but also meridional and other reflexions altered between rigor and relaxed insect flight muscle [3]. Furthermore, these alterations could be explained with the help of electron microscope images as arising from the coming and going of the strong 14.5 nm period structures on the thick (myosin) filaments and an apparent shape change of the myosin head between relaxed and rigor muscle. Here was a possible method for examining the shape of the macromolecules actin and myosin during contraction. These discoveries led to a period of active apparatus development culminating in 1970 in the use of synchrotron radiation as a source for X-ray diffraction.

As often happens new things arise from the fusion of two cultures: the macro-focus rotating X-ray generators at the Cavendish laboratory and at the Royal Institution were powerful devices based upon radiotransmitters as power supplies and a rotating anode built by Broad (in Cambridge) based on the design of Taylor. The cathodes were air-insulated and awesome. This was one ingredient in the then newly founded MRC Laboratory of Molecular Biology on Hills Road. The group from Birkbeck College which moved to the MRC lab in 1962 had another technical heritage. Under the influence of Rosalind Franklin, the Birkbeck group, to which I belonged, had taken over a French design of X-ray generator (Beaudouin) which used a cathode with a conventional high tension cable which, thanks to Vittorio Luzzatti, was equipped with a fine-focus electron gun. A shotgun wedding of the Broad anode with the Beaudouin cathode, regarded with great suspicion by Len Hayward the head of the workshop, was organised by Longley and myself. In the care of Tony Wollard, a technician of great skill who made many impossible things possible, this bastard

Topics in Current Chemistry, Vol. 151

became a great success and was soon to be taken over by Elliotts (later Marconi-GEC) as the first commercial fine-focus rotating-anode tube (GX6).

Fine focus was important for two reasons: for small widths the brightness attainable on a rotating anode is related inversely to the width so that it is very advantageous to make the source as small as possible. This criterion (which also applies for a stationary anode) had been exploited by Ehrenberg and Spear in the design of the microfocus Hilger and Watts X-ray generator, which had played an essential role in obtaining the diffraction patterns of DNA at Kings College. The other design criterion was that X-ray focusing elements, (essentially bent mirrors or bent monochromators) because of aberrations, cannot accept more than 50–100 μm of source at moderate distances (50 cm) from the source. DNA fibres are tiny (50 μm) and the whole apparatus had to be scaled accordingly. Smallness brings problems, however, mostly arising from fluorescence in the specimen, which the attrition of absorption and the inverse square law removes if one can stand back. Muscle offered larger specimens where the scale was often determined by the properties of the optical elements and the ability to utilise higher total X-ray flux. The GX6 filled the bill, and with a better electron gun developed by Elliotts, became the workhorse of muscle research for many years.

For studying weak scatterers such as muscle it is very advantageous to use focused monochromatic X-ray beams. Crystal monochromatisation reduces fluorescence by removing the hard X-ray component very effectively. Most devices used to focus X-rays are one-dimensional, relying on bending a plane surface to achieve focusing. The use of curved mirrors at glancing incidence for soft X-rays was pioneered by Franks at Birkbeck College (later at the NPL) [4]. Asymmetrically cut curved quartz monochromators were developed by Guinier and there was a strong French tradition with such devices which had been introduced to Birkbeck College by Rosalind Franklin, who had worked in Paris for some years. Both double-focusing mirrors (two mirrors at right angles) and tandem focusing monochromators at right angles were known and used, for example, by myself for TMV data collection. However, adjusting tandem monochromators at right angles is a tedious technique demanding excellent eyesight. Once again a fusion of two methods was successful. On the suggestion of H. E. Huxley, I set up a bent-mirror at right angles to a bent-monochromator. In combination with the GX6 the mirror-monochromator camera proved a very useful and robust tool for muscle research.

The technological thrust was then towards even brighter sources since even the GX6 was unable to support any but the simplest time resolved experiments. Heroic experiments by Huxley and Brown [5] managed to show the changes in X-ray diffraction from a frog muscle between the resting and activated states. Each of these experiments needed some hours of exposure time which had to be painstakingly collected with repeated stimulations of many muscles. In the mid-sixties I started wondering about synchrotron radiation. The theoretical work of Julius Schwinger [6] allowed one to calculate the spectral properties of synchrotron radiation. Such calculations showed that the DESY synchrotron, at 6 GeV then the most powerful source in Europe, would probably give an order of magnitude more monochromatic radiation than the best possible rotating anode tube. Madame Cauchois [7] had used a quarz monochromator with the Frascati machine to investigate the spectral properties in the X-ray region, but the electron energy at Frascati was too low to make this an interesting source for X-rays. The properties of the radiation were not in doubt, but the efficiencies of the focusing systems which would be used to collimate the beam were not well known. In 1966 I took up contact with DESY, specifically with Ruprecht Haensel, then setting up the F41 group to use DESY as an intense source of vacuum UV. In 1970, after moving to Heidelberg, I joined forces with Gerd Rosenbaum, a student in the F41 group, who came to Heidelberg to do his doctorate. Frustrations at the lack of enough intensity for time-resolved experiments on muscle even from improved rotating anode X-ray generators led to a renewed interest in the synchrotron which, in the meantime had increased its energy to 7.5 GeV, and its current to 20 mA. DESY agreed to sacrificing a few double shifts for our dubious experiments. Jean Witz was persuaded to leave Strassbourg for long enough to give his expert advice on X-ray optics. Conditions were not easy. In a double shift (16 h) a piece of the vacuum beamline had to be removed and an X-ray window installed. The optical bench then had to be installed and aligned with X-ray film (a great dicovery was that the beam was so intense that it turned microscope cover slips brown in a few seconds, this made life much easier!). A diffraction experiment could be caried out and, finally, the bench had to be removed and the beamline returned to its pristine condition before the end of the shift.

The first X-ray diffraction pattern with synchrotron radiation was obtained in September 1970 when a strip of insect flight muscle prepared by H-G. Mannherz was persuaded to give its equatorial diffraction pattern with curved quartz crystal monochromatised synchrotron radiation [8]. A rather smudged photograph demonstrated feasibility, and a new technology was born. All this happened in the F41 bunker at DESY, under the encouraging eye of Uli Arndt, one of the first converts to the new source. After our success we all retreated to a fish restaurant in Blankenese, and watched the great ships go by.

Much encouraged by this result, things started in earnest. The DESY directorium were very helpful and, in particular, Martin Teucher, who was Godfather to the project, gave the most real support possible by organising the building of a second synchrotron radiation bunker (Bunker 2) for X-rays. Rosenbaum was joined by John Barrington Leigh and later by Rolf Chors and Arnold Harmsen. Together they formed a small resident DFG-financed group. Jean Witz unfortunately did not stay with the team, perhaps because the French restaurant near DESY burnt down.

The performance of a synchrotron radiation source depends critically on the optical system used to bundle a tiny fraction of the total available energy through the specimen. For use with fibre specimens a monochromatic source is essential. However, fibres are not sensitive to the degree of convergence or divergence in the beam in the way crystal specimens are. Such considerations indicated to us that we should use a tried and trusted technology, namely the mirror-monochromator camera which had been developed in Cambridge. Rosenbaum and Barrington Leigh [9, 10] duly set up an elaborate remote controlled optical bench with three television cameras, one in line with the beam, for observing the beam and focus. The experimental area was shielded by a massive concrete wall. Because synchrotrons inject and dump their beam 50 times a second, there was a real possibility of the beam colliding with the walls of the ring. If this happened it would produce a neutron shower of some severity. The main beam shutter was substantial and could only be opened and closed from the main control room. Hence the complete remote control.

Compared with conventional X-ray sources, the scale of the apparatus was unusual. The focusing elements were a 40 cm segmented mirror nearest the synchrotron followed by a 10 cm bent quartz asymmetrically cut monochromator used in the anti-Guinier position, which leads to some wavelength inhomogeneity but has the advantage of compressing the beam. Specimen-film distances were 1–2 m. A segmented mirror was favoured because, to enable bending, the mirrors should be thin, whereas for stability and polishing they should be thick. The mirrors used were in fact too thin, so that after a year or so they became permanently deformed. Mirror thickness is an even now a parameter which gives concern. The first apparatus was partly designed and almost totally manufactured by H. Wagner, D. Müller, and W. Gebhard in the Max Planck Institute for Medical Research in Heidelberg. Rosenbaum and Barrington Leigh routinely achieved a focus of 200×500 µm, which was until quite recently the best focus obtained with synchrotron or storage ring sources. Their design became a model for other synchrotron radiation sources and is even now the most commonly used focusing system used for producing intense beams of (not strictly) monochromatic X-rays.

The Rosenbaum-Barrington Leigh system was not particularly easy to use. Remote control was powered by a hundred small DC motors, which led to an elegant light mechanical design but to a web of cables which were inclined to get damaged. Often two shifts would be taken to establish an optimal focus, by which time most of the experimental team were often too tired to do much with it. Most of the recording was done on X-ray film. However, another device of great utility at this time was the linear position sensitive detector developed by Gabriel and Dupont [11] in Grenoble, which enabled us to start time resolved experiments since the results could be stored in a pulse-height analyser. Andre Gabriel became a frequent visitor and later a staff member at Hamburg. Andre's apparatus used two FET's (field emission transistors), which were very sensitive to current overloads, to read out from the two ends of the counter. A hefty sneeze in the control area could encourage the FET's to burn out. Replacing FET's in the middle of the night after 72 h. of experiment was an experience one would rather not have had!

During this time the history of the synchrotron radiation group became entwined with the history of EMBL. We reported our early results to John Kendrew and to Hugh Huxley then advising on the setting up of EMBL. As a result of their enthusiasm it was decided to set up an Outstation of the EMBL laboratory at DESY. Bunker 2, duly enlarged with a second floor containing a workshop, some offices, and a chemistry laboratory, was to be taken over by EMBL. These were difficult moments since we had contracted for all sorts of things in the expectation of EMBL picking up the bill. However, EMBL itself was not ratified and therefore, legally, did not exist. At last the Italians (the last founder country to become a signatory) signed and we could, once more, sleep easy. The first move after the ratification was to employ Victor Renkwitz to organise the workshop. After ratification, the collaboration between DESY and EMBL was documented in an agreement which was duly signed with festivities in Jacobs Fährhaus by Sir John Kendrew as director general of EMBL and Prof. Schopper the director of DESY

Fig. 1. The signing of the agreement setting up the EMBL outstation. Shown are from the right to left Prof. Herwig Schopper, Director General of DESY, Prof. Sir John Kendrew, Director of EMBL and Heinz Berghaus, Director of Administration at DESY
The ceremony took place on April 21, 1975.

(Plate 1). This was the last time I remember seeing Martin Teucher, who was to die soon afterwards. A bitter loss.

Work in Bunker 2 in the early seventies was mostly concerned with muscle. Pioneering experiments were carried out by members of the Heidelberg Institute, particularly by Roger Goody, who carried out the first X-ray titration [12]. Richard Tregear, from Oxford, was a frequent visitor, and who now with encouragement and now with bullying, forced the pace of the early development. One of his most elegant experiments, namely a demonstration that a cyclic stretching and releasing of rigor insect flight muscle fibres has absolutely no effect on the diffraction pattern, has unfortunately never been published. Hugh Huxley also worked with us, but he soon established his own facility at Daresbury in collaboration with John Haselgrove and Uli Arndt. His return to Hamburg was to await the inauguration of the storage ring source (DORIS). One can summarise the early work on insect flight muscle by saying that the synchrotron radiation source, because of its laser-like collimation enabled us to record and measure accurately the diffraction pattern in various physiological states [13]. However, the DESY synchrotron was not bright enough to enable the time-resolved experiments with millisecond resolution we had all dreamed about. This had to await the storage ring (DORIS). Moreover, a study by Harmsen, Leberman, and Schulz [14] showed that the DESY source offered only marginal advantages over conventional sources for protein crystallography.

The end of the beginning came with the storage ring. It was clear that storage rings were in principle much better than synchrotrons. The DESY synchrotron held its own for so long because of its high energy. During the 1970s there was much activity in a number of electron storage ring facilities to set up X-ray diffraction facilities. These were established at Orsay, Stanford, Novosibirsk, Daresbury, Cornell, Brookhaven, Tsukuba, and in Hamburg (DORIS) and have proved enormously succesful for protein crystallography. The Doris facility was housed in

Bunker 4, a much grander affair than Bunker 2, which duly became the EMBL Outstation in Hamburg. In 1976, soon after the move to Bunker 4, I gave up the direction of the group which passed to Prof. Stuhrmann. Soon afterwards, our beloved optical bench in Bunker 2 was scrapped to make room for a white radiation (Laue) facility. This early attempt at what is now known to be a very powerful method was unfortunately not crowned with success, perhaps because of the unsuitable spectral properties of the synchrotron, which are biased to very hard radiation.

The rest of the present phase of development is a story of engineering and organisation now being carried out in numerous laboratories. Professional exploitation yields the rewards. Systems are very complicated and must increasingly be computer-controlled. The developments of detectors and computers capable of handling data at very high data-rates is a major preoccupation. For the future, undulators offer breathtaking new possibilities which will trigger new technologies. However, synchrotron radiation alone cannot provide all the answers. Muscle turns out to be more elusively complicated than we had imagined. The low angle scattering patterns we could obtain did not in fact have sufficient resolving power to tell us about the phenomenon of primary interest, namely what happens to a myosin crossbridge when a muscle contracts. While the method of X-ray scattering from intact muscle fibres remains our best hope for monitoring the dynamics of muscle contraction, it cannot be analysed without knowing much more about the structure of the components. Supplementary structural and kinetic information is necessary.

References

1. Huxley HE (1957) The double array of filaments in cross-striated muscle. J Biophys Biochem Cytol 3: 631
2. Huxley HE (1963) Electron microscope studies on the structure of natural and synthetic protein filaments from striated muscle. J Molec Biol 7: 281
3. Reedy MK, Holmes KC, Tregear RT (1965) Induced changes in the orientation of the cross bridge of glycerinated flight muscle. Nature 207: 1276
4. Franks A, Breakwell PR (1974) Developments in optically focusing reflectors for small angle x-ray scattering cameras. J Appl Cryst 7: 122
5. Huxley HE, Brown G (1967) The low angle x-ray diagram of vertebrate striated muscle and its behaviour during contraction and rigor. J Molec Biol 30: 383
6. Schwinger J (1949) On the classical radiation of accelerated electrons. Phys Rev 75: 1912
7. Cauchois Y, Bonnelle C, Missoni G (1963) Premiers spectres X du rayonnement d'orbite du synchrotron de Frascati. C r hebd Seanc Acad Sci Paris 257: 409
8. Rosenbaum G, Holmes KC, Witz J (1971) Synchrotron radiation as a source for X-ray diffraction. Nature 230: 434
9. Barrington Leigh J, Rosenbaum G (1974) A report on the application of synchrotron radiation to low-angle scattering. J Appl Cryst 7: 117
10. Barrington Leigh J, Rosenbaum G (1976) Synchrotron x-ray sources as a tool in biological structure and kinetic analysis. Ann Rev Biophy Bioeng 5: 239
11. Gabriel A, Dupont Y (1972) A position sensitive proportional detector for X-ray crystallography. Rev scient Instrum 43: 1600
12. Goody RS, Barrington Leigh J, Mannherz HG, Tregear RT, Rosenbaum G (1976) X-ray titration of binding of beta,gamma-imido-ATP to myosin in insect flight muscle. Nature 262: 613

13. Goody RS, Holmes KC, Mannherz HG, Barrington Leigh J, Rosenbaum G (1975) Cross bridge conformation as revealed by x-ray diffraction studies in insect flight muscle with ATP analogues. Biophysical J 15: 687
14. Harmson A, Leberman R, Schulz GE (1976) Comparison of protein crystal diffraction patterns and absolute intensities from synchrotron and conventional X-ray sources. J Molec Biol 104: 311

Applications of Synchrotron Radiation to the Study of Biopolymers in Solution: Time-Resolved X-Ray Scattering of Microtubule Self-Assembly and Oscillations

Eckhard Mandelkow, Gudrun Lange, and Eva-Maria Mandelkow

Max-Planck-Unit for Structural Molecular Biology c/o DESY, Notkestraße 85, D-2000 Hamburg 52, F.R.G.

Table of Contents

We have used synchrotron X-ray scattering from solutions to study the assembly of biopolymers. The principles of the method are reviewed, and applications are given for the example of microtubule self-assembly. We show how different structural intermediates of the reaction can be distinguished. Assembly can take the form of a first-order approach to a steady state, but non-equilibrium phenomena are also observed. An example is the unique behavior of synchronized oscillations in microtubule assembly.

Topics in Current Chemistry, Vol. 151

Eckhard Mandelkow, Gudrun Lange, and Eva-Maria Mandelkow

1 Introduction

Microtubules are fibrous structures that form part of the cytoskeletal network of eukaryotic cells. They are responsible for a variety of cellular functions, for example, the static support of a cell (axostyle), movement against the external medium (cilia: directed streaming; flagella: swimming), intracellular transport (e.g., in nerve cells), or the separation of chromosomes during mitosis. Microtubules are hollow cylinders, about 25 nm wide, built of the subunit protein tubulin. There are two forms of tubulin (α and β) of molecular weight 50 kDa. Together they form a heterodimer (α—β) which is the effective assembly unit. A linear string of α-β-heterodimers is called a protofilament; 13 of these associate laterally to form the microtubule cylinder (Fig. 1c). A variety of microtubule-associated proteins (MAPs) can be attached to the outside of the microtubule core. They tend to stabilize microtubules and enhance the efficiency of self-assembly. The mixtuture of tubulin and MAPs is called microtubule protein. Tubulin and MAPs can be isolated from mammalian brain and induced to form microtubules in vitro. This process requires binding and hydrolysis of the nucleotide GTP, and it can be controlled conveniently by temperature (microtubules form at 37 °C and fall apart at 4 °C). The protein can assemble into a variety of polymorphic forms, in particular ring-like oligomers that are prominent with microtubule protein at low temperature (Fig. 1a). The rings also contain bound MAPs, and therefore ring-containing solutions assemble more efficiently than purified tubulin. In the past several assembly models have been proposed in which rings were considered as nucleating centers (review [1]; the actual mechanism is different, as described below).

Microtubule assembly can be studied by different methods. The principal one has been UV light scattering or turbidimetry [2], others are viscosimetry, ultracentrifugation, or synchrotron X-ray scattering. One advantage of the latter technique is that it yields kinetic and structural information simultaneously, in contrast to others which lack one or the other [3–5]. These methods are sensitive to the average of all particles in solution. Since the particles are randomly oriented, the X-ray scattering can be analyzed by solution scattering theory. In cases where information on individual microtubules is required, one has to employ electron microscopy. In general these approaches are complementary to one another.

The X-ray studies on microtubules proceeded in several stages. The initial step was static X-ray fiber diffraction of oriented microtubules which yielded the main diffraction maxima and their structural interpretation [6]. Later, X-ray instruments, detectors, and data acquisition systems became available that were suitable for time-resolved X-ray diffraction of fibers (e.g., muscle) and solutions (reviews [7, 8]). All experiments described here were performed on the X-ray instruments of the EMBL outstation at DESY, Hamburg [8]. An initial study [3] showed that a solution of assembling microtubules could indeed be monitored by X-ray scattering, and that the scattering patterns could be interpreted on the basis of the oriented fiber pattern and additional information derived from electron microscopy and image reconstruction. The experiments required sample cells that were compatible with X-ray scattering (implying a small path length, about 1 mm, and windows transparent to X-rays, e.g., 50 µm mica) and yet allowing rapid and reversible temperature jumps up to 40 °C; this was achieved by a system of circulating water baths under feedback control [9]. Tubulin

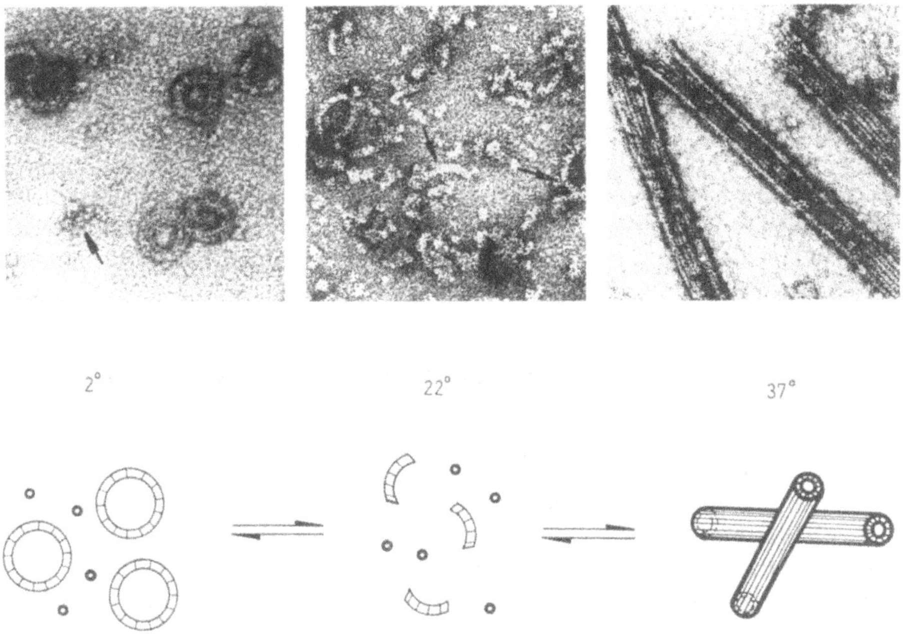

Fig. 1. Reaction scheme derived from X-ray scattering and electron micrographs of microtubule protein during assembly induced by a slow temperature scan. *Left*, rings of diameters around 36 nm an smaller oligomers at low temperature, negatively stained with 1% uranyl acetate. *Center*, breakdown of rings into smaller units at intermediate temperature, prior to microtubule assembly. *Right*, microtubules at 37 °C. Note the longitudinal protofilaments, spaced about 5 nm apart. Tubulin heterodimers are aligned along the protofilaments in a polar fashion (axial spacing of monomers is 4 nm). Magnification 200,000. Adapted from [3, 5]

can assemble not only into microtubules but also into oligomers that are often closed into rings. The rings played an important role in early assembly models, but there were conflicting views regarding their structure. The discrepancies could be resolved by X-ray scattering [10], showing that rings were equivalent to short coiled fragments of protofilaments. Subsequently we employed very slow temperature scans to define a series of structural transitions between cold microtubule protein (containing rings, dimers, and other oligomers) and microtubules at 37 °C (Fig. 1), [4, 5]. This led to the distinction between pre-nucleation events, nucleation, elongation, and post-assembly events. The kinetics of pre-nucleation events and microtubule assembly was subject of another study [11] whose main results will be described below. Finally, microtubule assembly has traditionally been described in terms of the theory of helical nucleation and condensation, in analogy with actin, this predicts a pseudo-first-order approach to a steady state of assembly [12, 13]. By contrast, microtubules can show distinct modes of non-steady-state assembly, including overshoots [14] and the synchronized oscillations to be described below [15, 16]. In this regard microtubules are unique among the self-assembling biopolymers.

2 Methods

2.1 Protein Preparation

The protein was obtained from porcine brain as described earlier [17]. X-ray experiments were performed either with microtubule protein (=tubulin + MAPs) or with purified tubulin (lacking MAPs); this protein is obtained by phosphocellulose chromatography and therefore referred to as PC-tubulin. Concentrations typically range from 5 to 50 mg/ml.

2.2 Time-Resolved X-Ray Scattering Using Synchrotron Radiation

The X-ray experiments were performed at the EMBL outstation at DESY, Hamburg, as described elsewhere [3–5], using instruments X13 or X33 equipped with position sensitive detectors (linear or quadrant detectors; for details see [8]. Assembly and disassembly were induced by temperature jumps between about 4 and 37 °C (half time 4 to 10 sec) [9]. Reciprocal spacings were calibrated with respect to the collagen reflections from a sample of cornea or tendon. The scattering curves were normalized with respect to the incident intensity measured by an ionization chamber just upstream from the specimen chamber.

2.3 Data Analysis and Model Calculations

The observed intensities were corrected for background (measured from a buffer filled cell), detector response, and variations in incident intensity due to the decay of the current in the storage ring. The scattering data were interpreted as described in detail elsewhere, starting either from the X-ray fiber pattern of oriented microtubules or from model calculations based on Debye's formula [4, 5, 10]. Figure 2 shows calculated scattering curves of several representative model structures, including microtubules, rings, protofilament fragments, and tubulin dimers. The curves are normalized to the same total protein concentration so that the forward scattering ($S = 0$, not shown in the graph) represents the degree of polymerization. Microtubules show the highest central scatter and several subsidiary maxima separated by clear minima. Rings have a lower central scatter and side maxima that are approximately out of phase with respect to microtubules. In both cases the positions of the first side maxima are close to $S = 1.22/D$, where D refers to the mean diameters of the respective hollow cylinders. Since microtubules have smaller diameters than rings, their subsidiary peaks occur at higher S values. By contrast, both tubulin dimers and short protofilament fragments show smoothly decaying scattering curves. In practice one usually observes a mixture of several types of aggregates, especially at low temperature where rings, oligomers, and dimers are in equilibrium. Note that the minimum preceding the first maximum of microtubules is particularly sensitive to the presence of oligomeric structures.

Because of the beam stop and parasitic scatter the lowest observable S values are around 0.01 nm^{-1}. In the following the intensity integrated between about 0.01 and

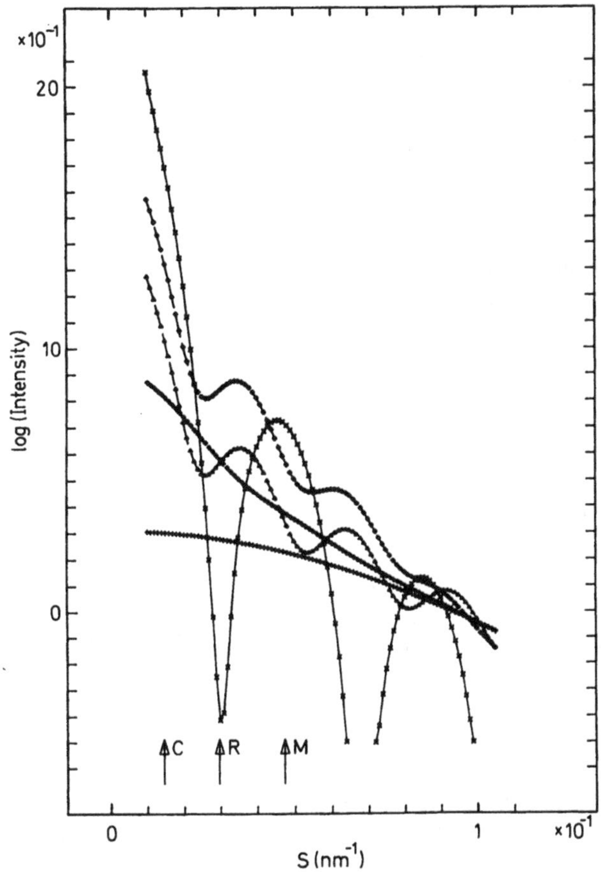

Fig. 2. Theoretical scattering curves of model strutures. The curves were calculated on the basis of Debye's formula, using spheres of 4 nm diameter to represent the protein subunits. The scattering of mixtures of aggregates can be calculated from $I(S, t) = \Sigma x_k(t) \cdot p_k \cdot i(S)$ where the index k refers to the species of aggregate $x_k(t)$ is the fraction of subunits assembled into the species k, p_k is the degree of polymerization, $i_k(S)$ is the form factor shown in the Figure, and $S = 2 \sin \theta/\text{lamda}$ is the Bragg scattering vector. From top to bottom on left-hand side: *Crosses*, microtubule with 13 protofilaments of mean diameter 26 nm (first subsidiary maximum at arrow M); *diamonds*, double concentric ring of 36 nm diameter (40 nm for outer, 32 nm for inner turn), with subunits spaced 4 nm apart (= coiled protofilaments; first side maximum at arrow R); *triangles*, single ring of 36 nm mean diameter (side maximum also near arrow R); *diamonds*, oligomer (= protofilament fragment consisting of four dimers; no pronounced side maximum); pulses, dimer of α and β tubulin. Note that the central scatter (*left*) increases with the degree of polymerization; the side maxima of rings and microtubules are roughly in antiphase; oligomers and dimers have smoothly decaying scattering curves, with the former contributing noticeably at small angles. The arrows marked C, R, and M indicate the regions sensitive to overall assembly ("central scatter", C), presence of rings (R) or microtubules (M). They are used to monitor the time-dependence of the reactions. From [11]

0.015 nm^{-1} will be referred to as "central scatter" for short (note that this is only an approximation to the forward scattering); integration between about 0.025 and 0.035 nm^{-1} yields the "ring scatter" since this overlaps with the first side maximum of the form factor of rings, and the region from 0.045 to 0.05 nm^{-1} will be called "microtubule scatter" (arrows C, R, and M in Fig. 2). The time-dependent changes in these regions allow one to asses the main features of assembly and disassembly. For example, a rise in the central scatter indicates overall assembly, a rise in the microtubule scatter usually means formation of microtubules, and a rise in the ring scatter may mean ring formation and/or other structures (e.g., oligomers). The actual structures present may be analyzed from the scattering curves or difference plots at given time points.

3 Results and Discussions

3.1 Prenucleation Events and Assembly

Figure 3 illustrates one cycle of assembly and disassembly of microtubule protein in standard buffer conditions. The projection plot shows the time dependence of the scattering traces. The initial state has a lower central scatter (left) and the side maxima typical of rings of diameter 36 nm; the polymerized state has a higher central scatter and the side maxima characteristic of microtubules (mean diameter 25 nm). A detailed comparison with the calculated scattering of model structures [4, 5] shows that most observed traces can only be explained if one assumes mixtures of several types of aggregates. At high temperatures the patterns are dominated by the scattering from microtubules, with a minor contribution from species that are best explained in terms of oligomers and/or MAPs that contribute mainly at very small scattering angles and fill in the minima of the microtubule trace. The scattering at low temperature can be modelled by mixtures of single and double concentric rings plus oligomers. The scattering from double rings has a high first subsidiary peak but decays more quickly than the scattering from single rings (see Fig. 2). As before, oligomeric species contribute mainly at small angles, with radii of gyration on the order of 10 nm (determined from the slope of Guinier plots). The contribution from tubulin dimers is generally small and thus negligible within noise until one reaches higher scattering angles (second side maximum and beyond, see Fig. 2). However, the presence of species with unknown form factors allows only an approximate quantitation of the individual components, especially at low temperature and during the assembly phase. When the temperature is raised to 37 °C there is a drop in intensity before the steep rise. This undershoot is due to the partial disappearance of ring oligomers before microtubules are formed (Fig. 1) and indicates that rings do not act as nucleating centers, as postulated in some earlier assembly models. However, in this experiment the phases of oligomer dissolution and microtubule formation are not well separated so that it is difficult to determine the kinetic parameters of each reaction separately.

Figure 4 shows the time courses of the intensities in the regions designated in Fig. 2 as central scatter (C), ring scatter (R), and microtubule scatter (M); the temperature

Fig. 3. Assembly of microtubule protein monitored by time-resolved X-ray scattering. The projection plot shows an experiment using the EMBL instrument X33, camera length 3 m, linear position-sensitive detector with 256 channels, 256 time frames of 3 sec per run (not all shown), temperature jumps from 3 to 37 °C and back (arrows). Note the increase in central scatter during assembly and the change in side maxima. The side maximum of the cold solution is due to rings, that of the warm solution is due to microtubules. From [11]

(T) is shown in the bottom trace. In this experiment the solution conditions were adjusted to obtain a better separation of prenucleation events and assembly (e.g., lower pH and salt concentration, see [11]). Following the temperature jump there is a pronounced undershoot (best visible in the top curve of Fig. 4) which appears to level out before the onset of the rise due to microtubule assembly. The undershoot follows nearly linearly the temperature rise; in other words, the observed rate does not reflect the intrinsic rate of oligomer dissolution, but rather a rapid temperature-dependent equilibrium. By contrast, microtubule assembly follows slowly, well after the temperature has reached 37 °C (for a more detailed analysis of this phase see [4, 5]. Other experiments show that the pre-nucleation phase occurs even in conditions where microtubule formation is inhibited, for example, in the absence of GTP, or

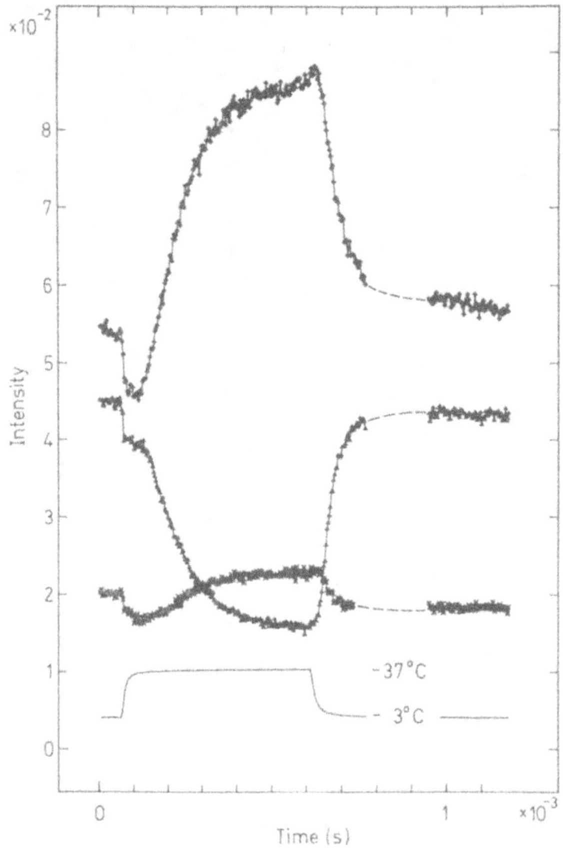

Fig. 4. Time-resolved assembly experiment, showing (from top to bottom) the time course of the intensities in the scattering regions C (*top*), R (*middle*), and *M* (*bottom*, compare Fig. 2); the solid trace below the experimental curves is the temperature T. Note the pronounced undershoot in the top curve accompanying the temperature jump, due to the disappearance of ring oligomers. The subsequent rise in the first and third curve is due to the nucleation and growth of microtubules. From [11]

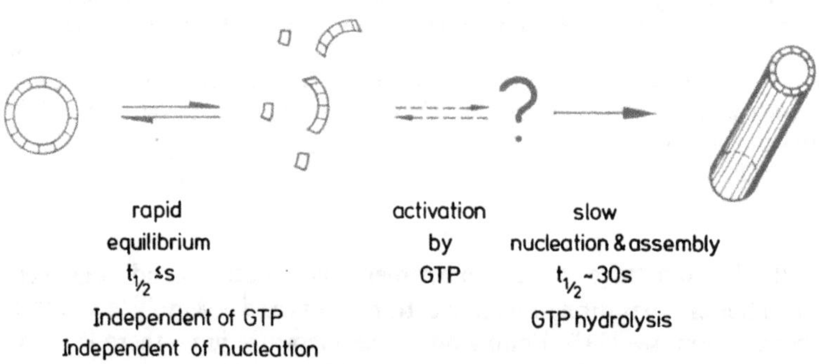

Fig. 5. Diagramm summarizing the differences between pre-nucleation events and microtubule assembly. There is a rapid and reversible temperature-dependent equilibrium between rings, smaller oligomers, and dimers that is not directly related to microtubule assembly (left part of diagram). By contrast, microtubule nucleation and assembly are show and take place above a threshold temperature (right part of diagram). Since the ring-forming and microtubule-forming modes of assembly draw on the same pool of building blocks (oligomers and dimers), one requires one or more intermediate steps during which the subunits are activated for assembly, e.g., by temperature and GTP. The structures of the activated dimers or oligomers are unknown but could correspond to the straight conformation of protofilaments, in contrast to the inactive coiled conformation. From [11]

when the temperature is jumped from the cold to less than 20 °C (not shown). The following conclusions can be drawn from these experiments [11]:

- The formation of oligomers is exothermic, that of microtubules is endothermic.
- The equilibrium between rings and smaller units (e.g., dimers) is rapid, while microtubule nucleation and elongation is slower.
- The prenucleation events occur over a wide temperature range and do not depend on GTP binding or hydrolysis, in contrast to microtubule assembly.

These results mean that tubulin is capable of two different modes of aggregation, leading to oligomers (e.g., at low temperature, without nucleotides) or polymers (e.g., at high temperature and with GTP). The two types of reaction have distinct physico-chemical characteristics and are thus not directly related to one another. However, they draw on the same pool of subunits, i.e., tubulin dimers. This is summarized in the scheme of Fig. 5.

3.2 Oscillations in Microtubule Assembly

The growth of microtubules, like that of other self-assembling biopolymers, is traditionally described by the endwise addition of subunits to the polymer at a roughly constant polymer number concentration [12, 13]. Thus, the elongation can be described by the reaction $MT_n + S < ... > MT_{n+1}$. This scheme predicts an exponential approach to a steady state where microtubules are in equilibrium with a critical concentration of free subunits. In standard assembly conditions this is indeed observed (see Figs. 3,4). However, microtubules are capable of more complex reaction mechanisms that are not compatible with the above scheme.

Several of these are illustrated in Fig. 6. Curve 1 was obtained with purified tubulin (largely devoid of the cold-stable oligomers because of the removal of MAPs). Its assembly shows the exponential approach to a steady state expected for simple endwise addition of subunits. Curve 2 shows an analogous experiment with microtubule protein (containing MAPs and therefore a larger fraction of cold-stable ring oligomers, similar to the experiments of Figs. 3 and 4). Here, too, one observes the exponential approach, apart from the undershoot due to ring dissolution preceding assembly (as described above). Curve 3 shows a rapid assembly phase which in this case is followed by a slow decay. This experiment was performed with a low concentration of GTP so that a large fraction of the nucleotide was consumed during the assembly phase. Since the maintenance of steady-state polymers requires free subunits with bound GTP, the observed slow decay can be explained by the lack of assembly-competent subunits so that the dissociation events are no longer balanced by association events. This is analogous to a single-turnover experiment in enzyme kinetics. The experiment illustrates that the tubulin · GTP complex must be regarded as the assembling unit. Curve 4 shows a further stage of complexity: assembly, decay, and recovery. In this case Mg^{2+}, rather than GTP, was strongly reduced by complexing it with EDTA. The behavior can be explained by considering that nucleotide binding is mediated by Mg^{2+}; in other words, the ternary complex of tubulin · Mg^{2+} · GTP is the assembly-competent subunit. Mg^{2+} · GTP is available in sufficient quantity for the initial assembly phase; after hydrolysis in the polymer, Mg^{2+} · GDP is locked non-exchangeably in the microtubules. The free tubulin subunits have a reduced

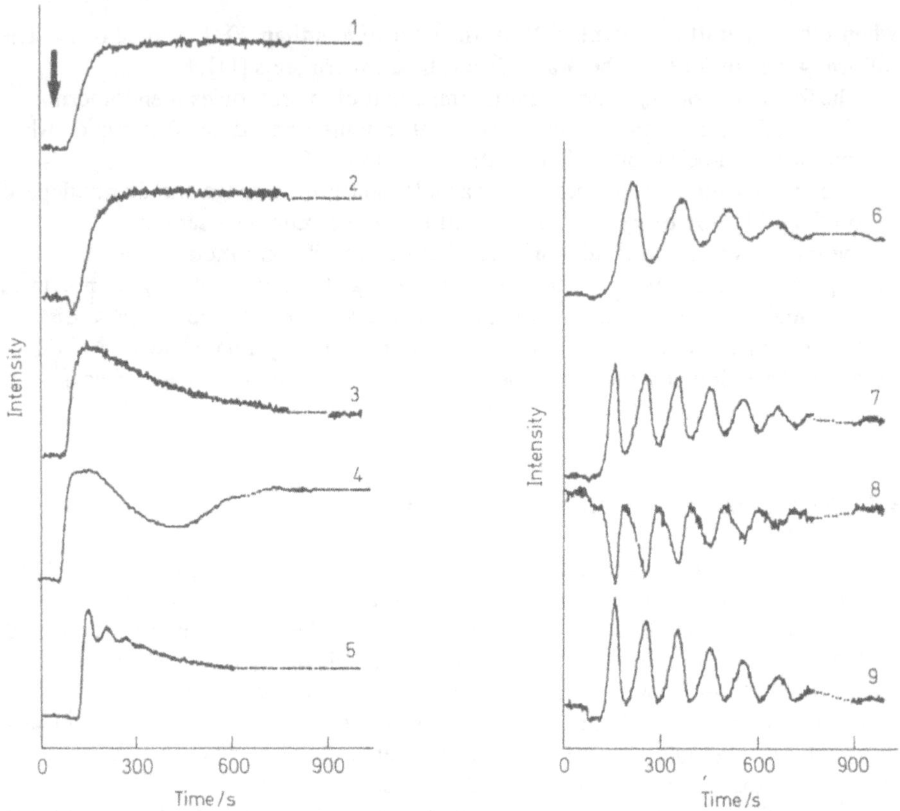

Fig. 6. Different types of microtubule assembly reactions followed by the time course of X-ray scattering at selected scattering angles. Curves 1 to 7 show the intensity at very low angles sensitive to overall assembly (region C, see Fig. 2), curve 8 is from a region sensitive mainly to rings and smaller oligomers (R in Fig. 2), curve 9 is sensitive to the formation of microtubules (M in Fig. 2). The arrow marks the temperature jump from 4 to 37 °C. The dotted regions represent breakds caused by data transfer to the computer. From [16]

Curve 1: Purified tubulin, concentration 35 mg/ml, in standard assembly buffer. The initial rise in itensity following the T-jump is due to nucleation followed by elongation, showing a first-order approach to a stable steady state.

Curve 2: Microtubule protein (containing tubulin and MAPs), 27 mg/ml, standard buffer. The curve is similar to the previous one, except that there is an undershoot preceding assembly (due to the dissolution of rings into smaller oligomers and subunits, similar to Fig. 3, 4).

Curve 3: PC-tubulin at low GTP/tubulin ratio (1 mM GTP, 36 mg/ml tubulin). The microtubule assembly phase is followed by a slow decay due to the depletion of GTP. The final state shows mainly oligomers.

Curve 4: Microtubule protein (17 mg/ml) in low Mg^{2+} concentration (2 mM $MgSO_4$, 4 mM EDTA). The initial rise and decay is similar to curve 3, in this case due to the transient depletion of Mg^{2+}. GTP; this is followed by a slow recovery of microtubule assembly to a new steady state.

Curve 5: PC-tubulin showing several oscillations at very high protein concentrations (46 mg/ml) in standard assembly buffer (compare curve 3).

Curve 6: PC-tubulin showing pronounced oscillations (periodicity 2.5 min) in oscillation buffer.

Curves 7–9, oscillating sample of microtubule protein monitored at three different scattering angles. Curve 7, region C (measuring overall assembly), curve 8, region R (oligomers), curve 9, region M (microtubules). The oscillations of curves 7 and 9 are roughly in phase because overall assembly (curve 7) correlates with microtubule assembly (curve 9). Curve 8 is roughly in antiphase because this scattering region is dominated by oligomers affinity for GTP in the virtual absence of Mg^{2+}, leading to the slow decay of microtubules similar to that of curve 3. After partial disassembly Mg^{2+} is released again from the subunits so that a new intermediate level of reassembly is attained.

The experiments described thus far set the scence for understanding the oscillations described below: Microtubules are intrinsically unstable after GTP hydrolysis; they can be stabilized by external factors (e.g., MAPs) or by continuously recharging the depolymerizing species Tubulin · GDP to the polymerizing species Tubulin · GTP. The recharging process Tubulin · GDP + GTP ↔ Tubulin · GTP + GDP is rapid with tubulin subunits so that a steady state can be effectively maintained as long as GTP is supplied; however, the direct recharging of subunits incorporated into microtubules or oligomers is not possible since the nucleotide exchange rates are nearly zero in these structures.

The non-equilibrium features can be strongly enhanced by appropriate choices of assembly conditions; the interval between consecutive assembly phases can be shortened, and the number of asembly cycles can be increased. Curve 5 of Fig. 6 was obtained at a particularly high protein concentrations in otherwise standard assembly buffer. Assembly becomes very rapid and runs into several damped oscillations before the system reaches a steady state at a lower degree of polymerization. A dramatic increase in oscillatory behavior is achieved by increasing the ionic strength ("oscillation buffer", Fig. 6, curves 6–9). The periodicities are typically in the range of

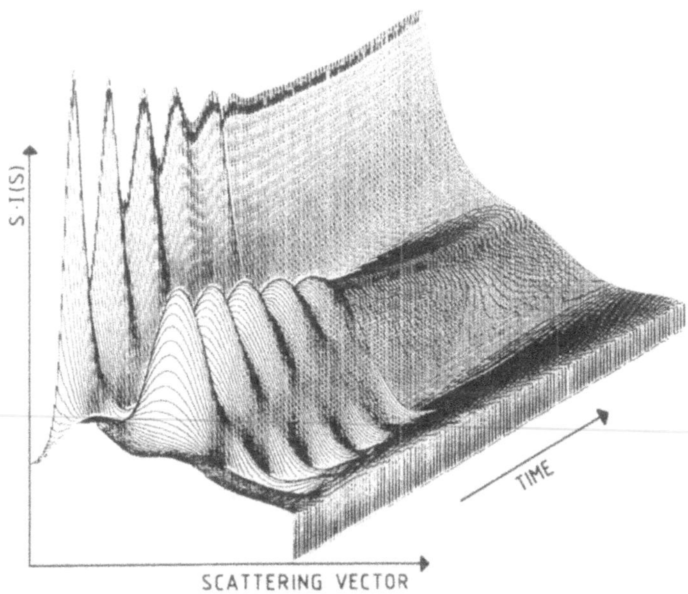

Fig. 7. Projection plot of the X-ray scattering from a solution of oscillating microtubules, showing the X-ray intensity (S · I(S), z-axis) as a function of scattering vector (S = 2 sin θ/lambda, x-axis) and time (y-axis, 3 sec scan interval). Microtubule protein, 32 mg/ml. The central scatter (*left*) indicates overall assembly, the subsidiary maximum arises from microtubules. The temperature jump is at time zero. The periodicity of the fluctuations is about 2 min. The final state (after disappearance of the oscillations) is dominated by the scattering from oligomers. The scattering curves here and in Fig. 8 have been smoothed by cubic splines. From [16]

1–3 min, and amplitudes can be up to 80% of the maximum. A projection plot obtained from an oscillating sample is shown in Fig. 7. The analysis of the scattering patterns (Fig. 8) shows that the reaction involves structures with distinct form factors so that different regions of the scattering pattern may swing either in phase or in antiphase (as in Fig. 6, curves 7–9). The oscillations may continue for up to 60 min; they correlate with the consumption of GTP and the buildup of GDP. The damping depends, among others, on the GDP/GTP ratio. Its increase leads to a lowered amplitude, with a relatively weak influence on the frequency.

What is the structural basis of the non-equilibrium behavior? We know from previous experiments that tubulin can exist in several states of aggregation that are interconvertible during assembly: tubulin dimers (the subunits of assembly), microtubules, and oligomers (short stretches of coiled protofilaments that tend to form closed rings when MAPs are present). Each of the structures has a characteristic X-ray pattern (Fig. 8a). When comparing the patterns at various time points during the oscillations one notes several nearly stationary isosbestic points (Fig. 8b),

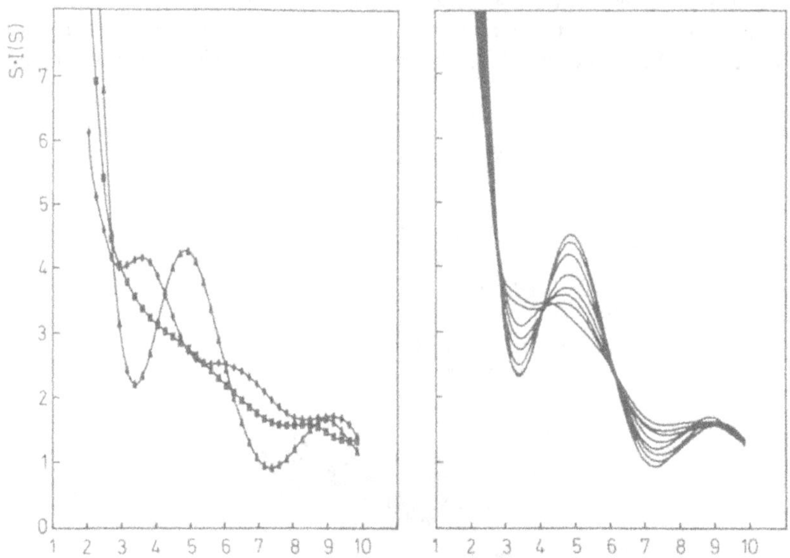

Fig. 8a, b. Scattering patterns of oscillating samples of PC-tubulin and microtubule protein. Vertical axis, I(S) · S (corresponding to the intensity integrated over circles of constant scattering angle 2θ); horizontal axis, reciprocal spacing S in mm^{-1}. **(a)** PC-tubulin, 29 mg/ml. *Crosses*, initial solution before T-jump, containing mainly tubulin dimers and a small fraction of oligomers. *Triangles*, microtubules at the first maximum of oscillations. Note the rise of the central scatter (*left*) and of the microtubule maximum around 0.05 nm^{-1}. *Rectangles*, final state after the oscillations are damped out, showing the smooth decay typical of non-ring oligomers. **(b)** Selected scattering traces observed during oscillations. The curve at the top of the central scatter (*left*) and at S = 0.05 nm^{-1} is at the first assembly maximum, the bottom curve is at the subsequent valley, the others are at intermediate time points. Note the constant positions of the isosbestic points where the curves intersect, indicating that the patterns are dominated by two main contributions (from microtubules and oligomers) whose ratios vary cyclically during the oscillations indicating that the patterns are dominated by two main species. One of them is microtubules, as seen from their characteristic side maxima. Subtraction of the microtubule component shows the second component to be oligomers (while tubulin subunits, the third major component, scatter only weakly and are thus disguised when the other structures are present). The amplitude of the oscillations represents the fraction of microtubule mass participating in the reaction; microtubule assembly is maximal at the peaks of the central scatter and minimal at the troughs.

This conclusion was confirmed by complementary negative stain electron microscopy experiments whose results are shown in Fig. 9. Although this method is less quantitative, it shows the features expected from the X-ray patterns, i.e., an antiphasic increase and decrease of microtubules and oligomers, typically 20 to 100 nm in length.

What is the origin of the oligomers? In principle they could form the disassembling subunits (containing bound GDP). However, since the nucleotide is exchangeable on free subunits, they would rapidly become assembly-competent again by binding of GTP from the solution. The alternative is that microtubules disassemble not via release of dimers, but via digomers. Direct evidence for this mechanism comes from earlier experiments where depolymerizing microtubule solutions in standard buffer were fixed by rapid freezing in amorphous ice and observed by cryo-electron microscopy [18]. Combining these results with the earlier observation

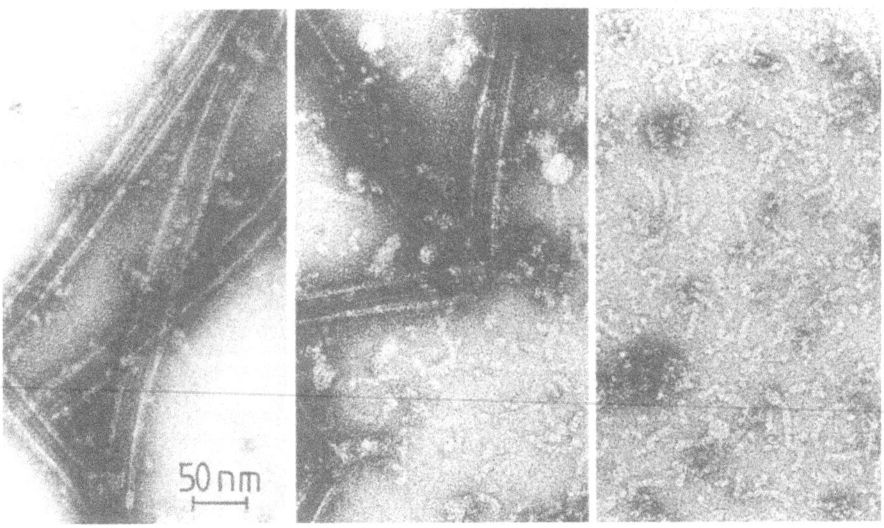

Fig. 9. Electron micrographs of PC-tubulin at different stages of oscillations. *Left*, maximal assembly. One observes microtubules of normal appearance and oligomeric material in the background. *Center*, phase of disassembly. There is a noticeable increase in oligomers, and microtubules tend to break. *Right*, field of oligomers at an assembly minimum. Typical lengths are between 20 and 100 nm, corresponding to 3 to 12 tubulin dimers. The longer ones clearly show the coiling characteristic of diassembled protofilament fragments that leads up to ring-like closure (see upper left). Complete rings are rare with PC-tubulin but frequent with microtubule protein due to the stabilization of oligomers by MAPs. From [15]

that oligomers tend to break down prior to assembly (Fig. 5) we arrive at the following simple model for the cyclic part of the reaction:

Subunits → Mikrotubules → Oligomers
(Oligomers → Subunits, return arrow)

Note that when the conversion from oligomers to subunits is rapid this reduces to the usual scheme of endwise assembly and disassembly, Subunits ↔ Microtubules.

Any chemical oscillator requires an energy source which in the present case comes from GTP hydrolysis (data not shown; see [15]: (a) Non-hydrolyzable GTP analogs support assembly but not oscillations; (b) the number of oscillations is limited by GTP, i.e., oscillations do not outlast the GTP supply; (c) microtubules are formed with GTP and destabilized by GDP, whereas oligomers are stable with GDP. Thus, one expects that the cyclic change of structures is accompanied by a cyclic change in the interaction between tubulin and GTP or GDP. This can be shown by analyzing oscillating solutions for protein-bound nucleotides by HPLC [16]. Both the levels of GTP and GDP undergo oscillations with the same frequency as that of assembly, but in antiphase; when GTP shows a peak, GDP has a trough. Thus, the energy of hydrolysis is not consumed continuously, but in bursts coupled to assembly. Control experiments showed that neither microtubules nor oligomers, but only tubulin subunits, were capable of exchanging nucleotides with the solution, and that the exchange on subunits is rapid.

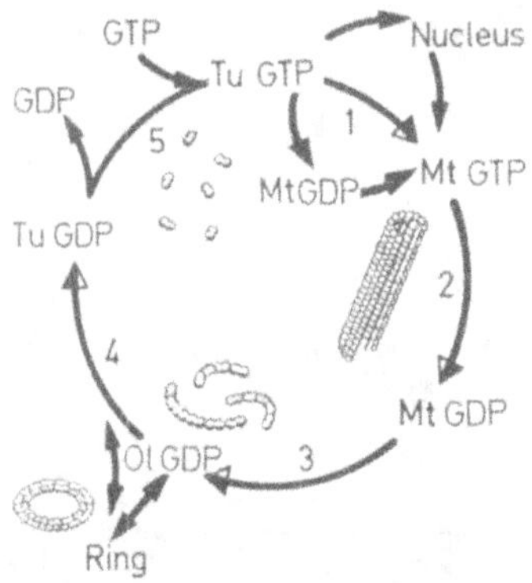

Fig. 10. Model of reaction cycle responsible for oscillations. Microtubules (Mt · GTP) are formed from active subunits (loaded with GTP, Tu · GTP, step 1), GTP is hydrolyzed upon incorporation of the subunits (Mt · GDP, step 2), leading to the destabilization of microtubules and eventually to their disassembly into oligomers loaded with GDP (Ol · GDP) which transiently lock the subunits in an unpolymerizable state. When inactive subunits (Tu · GDP) are released from oligomers (step 4) they can be recharged to Tu · GTP (step 5), leading to the reassembly of microtubules. Side reactions are the dissolution of rings into oligomers and subunits (observed with microtubule protein just after the T-jump and responsible for the undershoot, see Fig. 3, 4), and the nucleation during the first round of assembly. From [15]

These data are incorporated into the reaction mechanism of Fig. 10. From a structural point of view there are three species, tubulin subunits (Tu), microtubules (Mt), and oligomers (Ol), that are cyclically interconverted and can be distinguished by X-ray scattering. Functionally, the number is at least five because tubulin and microtubules come in two forms, either with GTP or GDP bound. GTP-charged tubulin (Tu · GTP) is assembly-competent and in equilibrium with microtubule ends; tubulin with bound GDP (Tu · GDP) normally cannot assemble and is in equilibrium with oligomers. Free tubulin subunits quickly exchange their nucleotides according to the GTP/GDP ratio in solution, microtubules and oligomers do not exchange nucleotides. Microtubules are initially formed with bound GTP (Mt · GTP) which is then hydrolyzed (Mt · GDP). Microtubules are destabilized by GTP hydrolysis; oligomers with bound GDP (Ol · GDP) break off and subsequently dissociate into Tu · GDP subunits which can then be recharged to Tu · GTP. Additional pathways listed in the scheme but less important for this discussion are (a) the prenucleation events observed mainly with microtubule protein, rings $\twoheadrightarrow$ oligomers $\twoheadrightarrow$ Tubulin · GDP (Fig. 5); (b) nucleation during the first round of assembly, Tubulin · GTP $\twoheadrightarrow$ nuclei $\twoheadrightarrow$ microtubules (subsequent cycles do not require new nucleation because microtubule ends are already present). This scheme can be applied to "classical" models of helical condensation as well as to oscillations.

The crucial point for oscillations is that oligomers, the primary breakdown products of microtubules, have a measurable lifetime (i.e., the rate of step 4 is slow). This means that the reactivation of tubulin subunits is temporarily interrupted because oligomers act as a temporary storage form of nonpolymerizable tubulin. Since the pool of polymerizable Tubulin · GTP is nearly depleted during the assembly process (step 1) and cannot be replenished quickly, the destabilization of microtubules due to GTP hydrolysis (step 2) becomes noticeable. The depolymerization (step 3) proceeds until the oligomers have released enough subunits to stop it and return to a new phase of assembly. GTP plays a dual role; its binding to subunits activates them for assembly, and its hydrolysis within the polymer induces their breakdown. In other words, GTP acts as an activator of the polymer, GDP activates the oligomer.

The reaction scheme of Fig. 10 suggests that the conditions for obtaining oscillations are not unique but rather depend on the relative magnitudes of several rate constants. For example, one would expect steady-state microtubules in conditions where microtubules are strongly stabilized, and steady-state oligomers where these are stabilized. These conditions are often mutually antagonistic since microtubules consist of straight protofilaments, whereas those of protofilaments are coiled. Experimental examples are illustrated in Fig. 11: All samples contain additional salt which destabilizes microtubules and — to a lesser extent — oligomers as well. In Fig. 11a the microtubules were stabilized by glycerol, leading to their rapid assembly after the T-jump and a steady state. When the glycerol is omitted (Fig. 11c) one observes a very slow intensity rise, due mainly to oligomers and a small fraction of microtubules, but also no oscillations. However, at an intermediate ratio of salt and glycerol one obtains oscillations (Fig. 11b).

The principles underlying these observations are summarized in the "pha diagram of Fig. 12. The amplitude of oscillations is plotted as a function of the ratio of stabilizing and destabilizing agents (for details see [16]). When microtubules are stabilized (left part of diagram) one observes microtubules without oscillations.

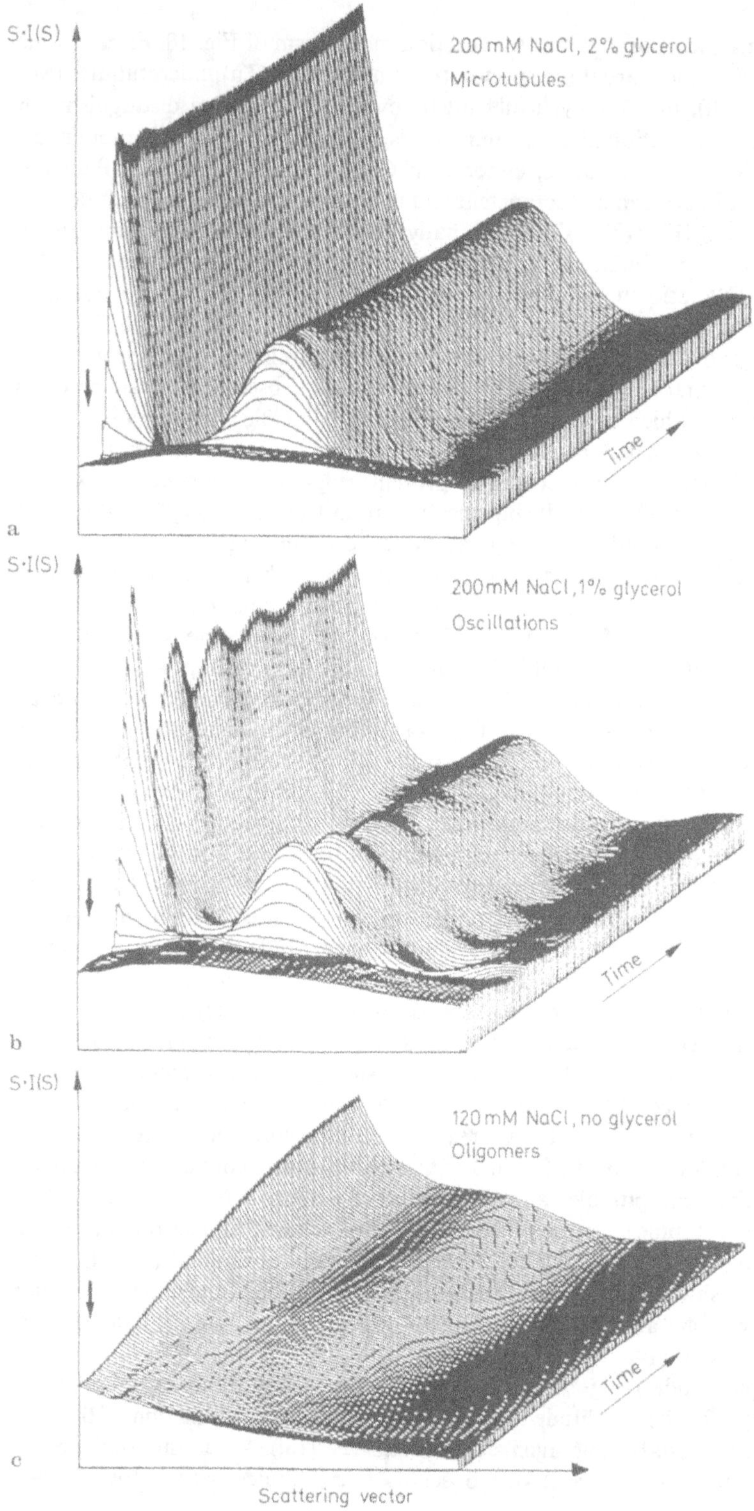
S·I(S)
200 mM NaCl, 2% glycerol
Microtubules
Time
a
S·I(S)
200 mM NaCl, 1% glycerol
Oscillations
Time
b
S·I(S)
120 mM NaCl, no glycerol
Oligomers
Time
c
Scattering vector

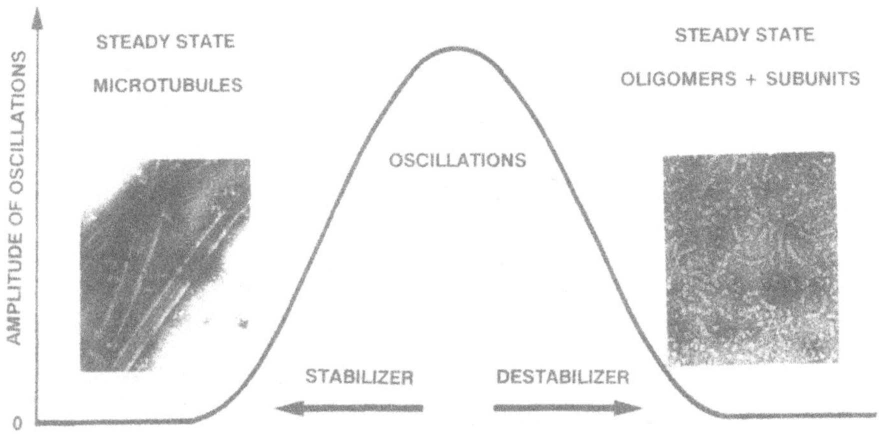

Fig. 12. Dependence of oscillation on the balance between stabilizers and destabilizers. The ratio of destabilizer to stabilizer is plotted along the x-axis, the magnitude of oscillations on the y-axis. On the right side of the diagram assembly leads to oligomers (excess of destabilizer), on the left side to microtubules (excess of stabilizer), in both cases without oscillations. Oscillations can be generated at intermediate ratios. Electron micrographs of microtubules and oligomers are shown as insets. From [16]

When oligomers are stabilized (right part of diagram) there will be oligomers without oscillations. Intermediate conditions (center of diagram) will lead to oscillations between microtubules and oligomers.

The check the validity of the arguments a series of model calculations based on the reaction mechanism of Fig. 10 were performed [19], details to be published). They are based on a set of coupled differential equations and on the assumption that the phase transition between microtubule growth and shrinkage can be explained by cooperative interactions between tubulin subunits at microtubule ends. Two examples are shown in Fig. 13 where the rate of GTP hydrolysis following the incorporation of tubulin into microtubules was varied. When this rate is fast (Fig. 13a) one finds pronounced oscillations (this is equivalent to an intermediate stability of microtubules, compare Fig. 11b). When the rate of hydrolysis is reduced (Fig. 13b) the oscillations disappear because microtubules are effectively stabilized, and they remain assembled in a steady state (compare Fig. 11a).

◂ **Fig. 11a–c.** Projections plots showing the action of microtubule stabilizers and destabilizers. Tubulin concentration 44 mg/ml, concentrations of NaCl and glycerol as below. **(a)** 200 mM NaCl, 2% glycerol. Although NaCl inhibits assembly (see below), the stabilization by glycerol dominates, leading to rapid microtubule formation and steady state (after some minor oscillations). **(b)** 200 mM NaCl, 1% glycerol. In these conditions the sample shows clear oscillations. **(c)** 120 mM NaCl, no glycerol. Microtubule formation is strongly inhibited by the salt, as seen from the weakness of the subsidiary maximum. Most of the increase in the central scatter (*left*) of from oligomers and higher aggregates

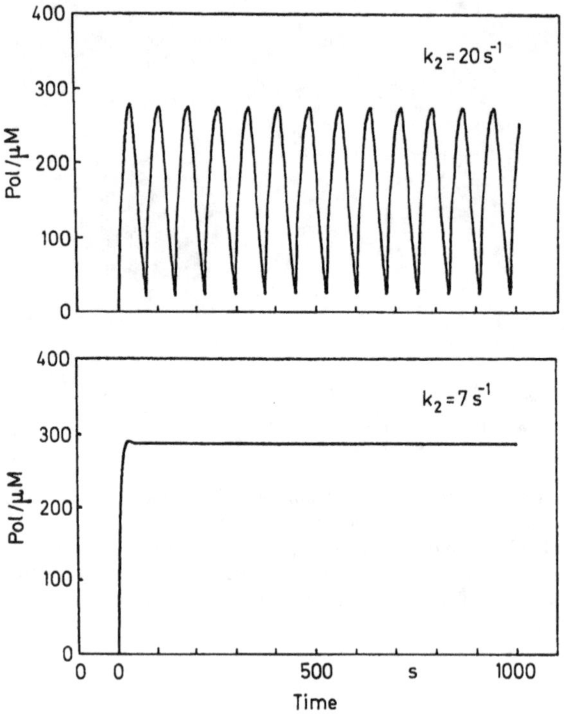

Fig. 13. Some representative model calculations of oscillations based on the reaction cycle of Fig. 10. Abscissa, time; ordinate, extent of polymerization. The first assembly is initiated at time zero (equivalent to a temperature jump). The two curves differ by the rate of GTP hydrolysis (k_2). If this rate is fast, the oscillations are pronounced (*top trace*); if it is slow, microtubules assemble to steady state. For details see [19]

3.3 Comparison with Chemical Oscillations and Implications for Microtubule Dynamics in Cells

Several examples of chemical oscillators are known, for example, the Belousov-Zhabotinskii reaction or anaerobic glycolysis in yeast [20] and it is interesting to note some parallels and differences with the present one. Firstly, they depend on a rechargeable supply of energy; in this regard they differ from a mechanical pendulum which cyclically converts one form of energy into another (neglecting friction). The loss of energy in chemical systems is explained by irreversible thermodynamics; basically it stems from the fact that they are based on many particles that act in concert against entropy, rather than a single oscillator without entropy change. Secondly, biochemical oscillators are usually composed of a series of enzyme reactions, some of which are allosterically regulated by reaction intermediates or products, and this feedback leads to nonlinear behavior. In our case the enzyme analog is the tubulin subunit whose polymerizability is regulated by binding of GTP or GDP, but tubulin does not function as an enzyme in the strict sense since hydrolysis occurs in the microtubule. Thus, the buildup and breakdown of a structure corresponds to the rise and fall of enzyme activities in other systems.

The oscillatory behavior of microtubules in solution is related to their dynamic instability [21, 22]. During the cell cycle microtubules must undergo a cyclic reorganization between different types of networks (e.g., the cytoskeleton of interphase and the mitotic spindle) which involves the periodic disassembly and reassembly. Moreover, even local regions of cells show microtubules switching between phases of growth and shrinkage. In these cases their behavior is not synchronized, and an observer recording the average degree of assembly would see an apparent steady state. In the experiments described here the whole microtubule population was synchronized by the rapid initiation of assembly. Although our solution conditions differ from the environment within a cell, the overall behavior is surprisingly similar (e.g., amplitudes and periodicities). In other words, it seems that the oscillations represent an amplified version of reactions that take place inside living cells. This opens the opportunity to study their basic properties in vitro.

4 Summary: Pros and Cons of Using Synchrotron Radiation for Studying Biopolymer Assembly

In this article we have concentrated on the uses of X-ray scattering from synchrotron radiation for the study of microtubule polymerization. Historically the main tool for monitoring microtubule assembly has been UV light scattering or turbidimetry [2], and this method is also capable of revealing the oscillations [23, 24]. The method is sensitive, easy to use, and available in most laboratories. It is therefore appropriate to ask what additional information one can gain from time-resolved X-ray scattering.

The disadvantages of X-ray experiments are evident: Time-resolved studies can only be done with synchrotron radiation because even the best conventional X-ray generators do not provide sufficient intensity. Synchrotron experiments are expensive, beam time is scarce, and there is the risk of beam damage. Because of the high costs, facilities have been set up in several countries which provide the radiation and the related instrumentation for a large number of visitors (e.g., the Hamburg Synchrotron Laboratory HASYLAB and the EMBL outstation at DESY, Hamburg; for an overview of the experimental facilities see the annual reports of these laboratories). This reduces the cost per experiment, and in addition it makes the techniques accessibles to users who do not have X-ray equipment in their own laboratories. The limited beam-time available to each user means that it is sometimes difficult to carry out extended sets of experiments so that a careful selection of research priorities is required. Fortunately, the problem of radiation damage turned out to be less severe than had been anticipated, and in many cases the measurements can be completed before damage becomes noticeable. The reasons is that at high dose rates the specific damage per incident photon is reduced (for a discussion see [4]).

The advantages of X-ray scattering can be summarized as follows:

(a) X-rays have a much shorter wavelength than UV light (1.5 Å in our experiments), and therefore they offer a superior resolving power. Thus, one can distinguish different types of structures participating in a reaction, for example, polymers and oligomers. Roughly speaking, the light scattering signal corresponds to the very low angle part of the X-ray trace. As illustrated in Fig. 2, this region (C) senses the degree of polymerization, but the structural information contained in the higher angle pattern is accessible only with X-rays.

(b) Because of the high intensity of synchrotron X-rays, one can obtain structural as well as kinetic information at the same time and from the same sample. This allows one to characterize the reaction mechanism in detail. The above two arguments (a) and (b) are probably the most obvious ones in favor of time-resolved X-ray scattering.
(c) While electron microscopy reveals the structure of single particles, X-rays record the weight average of an ensemble of particles. Therefore X-ray patterns give a statistically more reliable representation of a reaction. On the other hand, they are also more difficult to interpret. To make optimal use of the X-ray patterns one usually requires additional information, e.g., in the form of electron micrographs (single particles), oriented X-ray fiber patterns (yielding higher resolution than the solution scattering patterns from randomly oriented particles), or structural model calculations (the usual approach to the interpretation of solution scattering, employing for example Debye's formula).
(d) An important aspect of solution X-ray scattering is that the particles are in their native aqueous environment, and that the solvent parameters can be easily varied. This is usually not the case with electron microscopy which suffers from artefacts of dehydration and staining. Thus, in principle the X-ray data are more trustworthy.
(e) Finally, one feature of the X-ray data collection is that they are stored in a computer and therefore immediately accessible to digital data processing. This facilitates the comparison of a large number of experiments. By comparison, light scattering experiments usually yield a trace on a chart recorder which makes quantitative analysis cumbersome. Although this is only an ancillary bonus of the X-ray experiments it turns out to be important in practice. It allows one to relate different experimental conditions to one another, get immediate feedback on the results during the experiment, and thus use the valuable beam time efficiently.

Future developments: While the earlier synchrotron radiation laboratories emerged as side branches of high energy physics facilities there is now a new generation of synchrotrons dedicated exclusively to the production of synchrotron light (see article by C. Riekel, this volume). In addition, insertion devices such as wigglers or undulators are capable of increasing the intensity further by several orders of magnitude. This means that the irradiation of a sample will no longer be a limiting factor; rather, the limit will be in the speed at which scattering data can be recorded (e.g., by position-sensitive detectors or other devices) and stored in the computer. It would be desirable to reduce the sample volume or its concentration (especially for rare proteins), increase the time resolution (for fast reactions) and the spatial resolution (for higher structural detail), improve the signal-to-noise ratio even with weakly scattering objects, and record intensities over a wide dynamic range. These requirements tend to be mutually exclusive, but if the technical problems can be overcome it will be possible to perform new types of experiments.

Acknowledgements: We are grateful to Dr. Michel Koch and the staff of the EMBL outstation at DESY, Hamburg, for making the X-ray facilities available. This project was supported by the Bundesministerium für Forschung und Technologie and the Deutsche Forschungsgemeinschaft.

5 References

1. Kirschner MW (1978) Int Rev Cytol 54: 1
2. Gaskin F, Cantor CR, Shelanski ML (1974) J Mol Biol 89: 737
3. Mandelkow E-M, Harmsen A, Mandelkow E, Bordas J (1980) Nature 287: 595
4. Bordas J, Mandelkow E (1983) In: Shaafi RI, Fernandez SM (eds) Fast methods in physical biochemistry and cell biology, Elsevier/North Holland Biomedical
5. Bordas J, Mandelkow E-M, Mandelkow E (1983) J Mol Biol 164: 89–135
6. Mandelkow E, Thomas J, Cohen C (1977) Proc Natl Acad Sci USA 74: 3370
7. Rosenbaum G, Holmes KC (1980) In: Winick H, Doniach S (eds) Synchrotron radiation research. Plenum, New York, p 533
8. Koch MHJ, Bordas J (1983) Nucl Instrum Meth 208: 461
9. Renner W, Mandelkow E-M, Mandelkow E, Bordas J (1983) Nucl Instr Meth 208: 535
10. Mandelkow E, Mandelkow EM, Bordas J (1983) J Mol Biol 167: 179
11. Spann U, Renner W, Mandelkow E-M, Bordas J, Mandelkow E (1987) Biochem 26: 1123
12. Oosawa F, Asakura S (1975) Thermodynamics of the polymerisation of protein, Academic, London
13. Johnson KA, Borisy GG (1977) J Mol Biol 117: 1
14. Mandelkow E, Mandelkow E-M, Bordas J (1983) TIBS 8: 374
15. Mandelkow E-M, Lange G, Jagla A, Spann U, Mandelkow E (1988) EMBO J 7: 357
16. Lange G, Mandelkow E-M, Jagla A, Mandelkow E (1988) Eur J Biochem 178: 61
17. Mandelkow E-M, Herrmann M, Rühl U (1985) J Mol Biol 185: 311
18. Mandelkow E-M, Mandelkow E (1985) J Mol Biol 181: 123
19. Lange G (1989) Doctoral Thesis, University of Hamburg
20. Hess B, Boiteux A (1971) Ann Rev Biochem 40: 237
21. Mitchison T, Kirschner M (1984) Nature 312: 237
22. Horio T, Hotani H (1986) Nature 321: 605
23. Pirollet F; Job D, Margolis RL, Garel JR (1987) EMBO J 6: 3247
24. Carlier MF, Melki R, Pantaloni D, Hill TL and Chen Y (1987) Proc Natl Acad Sci USA 84: 5257–5261

5 References

High Angle Fiber Diffraction Studies on Conformational Transitions in DNA Using Synchrotron Radiation

Robert J. Greenall [1,2,*] and Watson Fuller[1]

1 Department of Physics, University of Keele, Staffordshire, ST5 5BG, U.K.
2 SERC Daresbury Laboratory, Warrington, Cheshire, WA4 4AD, U.K.

* Present address: Department of Physics, University of York, Heslington, York, YO1 5DD, U.K.

Robert J. Greenall and Watson Fuller

X-ray fiber diffraction techniques have shown that DNA and synthetic polynucleotides can adopt a variety of stable double-helical conformations. The conformations available to a particular polynucleotide depend upon the type and concentration of the cations surrounding the DNA molecule, the sequence of Watson-Crick base-pairs within the molecule and the amount of water in the fiber. Transitions between the available conformations may be induced by modifying the aqueous environment of the DNA molecule through varying the ambient relative humidity. The advent of high brightness synchrotron X-ray sources has allowed time-resolved diffraction studies to be undertaken as the DNA molecule changes from one conformation to another in response to the changing aqueous environment. Such conformational transitions may be of importance in biological control processes which are mediated by specific DNA-protein recognition.

We review the methods and theory of fiber diffraction and the instrumentation required for time-resolved studies at a synchrotron source. The conformational polymorphism of DNA is described and a detailed account is given of time-resolved X-ray studies on one particular transition. We conclude with some comments on future developments in this technique.

1 Introduction

X-ray high angle fiber diffraction has proved to be the most powerful and versatile technique available for the study of the structure of fibrous polymers. In general the secondary structures of fibrous molecules are helices which show a marked disparity in the magnitudes of their linear dimensions parallel and perpendicular to the molecular axis and they do not fold into the compact globular tertiary structures which are favourable for the formation of macroscopic single crystals. Hence they are not amenable to the three-dimensional crystallographic techniques employed for example in the study of globular proteins. They can, however, form into limited regions of full three-dimensional crystallinity often with the polymer snaking via amorphous regions between different microcrystallites. In these microcrystallites the molecular helix axis is usually parallel to the c-axis. While such microcrystallites have not been isolated for individual study, a well-oriented crystalline fiber contains many microcrystals whose c-axes are aligned so that they are closely parallel to the length of the fiber. In general, the microcrystallite azimuthal orientations are randomly distributed over 2π radians and there is no correlation between the positions of the microcrystallites within the volume of the fibre. As a result, the diffraction from each of the microcrystallites adds incoherently and, in effect, the pattern observed is similar in character to one collected from a small single crystal using the rotation method. Although there is usually some loss in structural information arising from the overlapping in the diffraction pattern of Bragg reflections with the same radial distance from the axis of rotation in reciprocal space (the meridian), the information that is obtained is nevertheless three-dimensional and can be used to derive a structural model of the molecule. The Bragg reflections fall onto layer-lines which are perpendicular to the meridian and whose separation is inversely proportional to the pitch of the helix. Since a helical molecule consists of pseudo-identical subunits in equivalent positions, the distance between the meridional reflections is inversely proportional to the axial separation of the subunits. Hence even a cursory inspection of the diffraction pattern is often sufficient to define the gross helical parameters of the molecule such as its pitch and the number of subunits in one turn of the helix. This information, in conjunction with constraints on stereochemistry provided by known details of covalent bond lengths and angles, can be used to define models which can be refined against the full set of diffraction data. In less favourable cases the fiber consists of an aggregate of molecules which is not crystalline but in which the helical axes are again roughly parallel to the fiber axis with random azimuthal orientation. Under these circumstances the diffraction pattern corresponds essentially to the rotation pattern from a single molecule. Although the diffraction along the layer-lines in such a pattern is continuous, the separation between layer-lines is still inversely related to the pitch of the molecule and the distance between meridional reflections is still inversely related to the axial spacing between adjacent helical subunits in the molecule.

The fiber diffraction technique has been used to determine the structures of a wide variety of synthetic and biological molecules including structural proteins such as collagen and keratin, a range of helical conformations of the nucleic acids, DNA and RNA, and polysaccharides. In the case of the B form of DNA, early fiber diffraction patterns, which were not fully crystalline, still provided sufficient information to show that the structure was helical with a pitch of 34 Å and 10 subunit repeats per helix pitch [1–3]. Parallel studies on the crystalline A form showed, on the basis of the most likely spacegroup designation, that the helix contained two antiparallel polynucleotide strands.

It was found that the diffraction patterns observed from the A and B forms of DNA were essentially independent of the biological origin of the DNA [4] and hence of the sequence of bases, adenine, thymine, guanine and cytosine, along each polynucleotide chain which Watson and Crick proposed as the structural basis for storing the genetic information. Since the Watson-Crick base-pairing scheme required that the base-pairs be on the inside of the molecule while the sugar-phosphate backbone be on the outside, this model immediately explained the paradoxical observation that a molecule containing a random sequence of bases could form into highly regular crystals. Intensive studies of naturally occurring DNA soon revealed that DNA molecules from a wide variety of biological sources could all adopt three major conformations designated A [5], B [6] and C [7] which differed from each other in terms of helical parameters. It was also observed that molecules within a fiber could undergo reversible transitions between these conformations, and in particular between the A and B forms. Three factors have been identified as important in determining the molecular conformation: (1) the nature of the cation neutralizing the negatively charged phosphate groups of the DNA together with the presence of other ions in the environment of the DNA [8]; (2) the degree of hydration of the DNA which is influenced by the relative humidity of the fiber environment [9]; (3) the base composition and sequence of the DNA (e.g. [10]). These effects, and particularly those associated with base composition, have been shown to be most marked for synthetic polynucleotides with a highly repetitive base sequence [11]. In addition to the A, B and C conformations, synthetic polynucleotides can adopt two conformations not observed for naturally occurring DNAs; these have been designated D [12] and S [13].

Since RNA double helices and most RNA/DNA hybrids have only been observed in conformations similar to the A-form [14–24] whereas studies of the DNA conformation in cells such as salmon sperm have revealed that the "resting" state of DNA appears to be the B-form [25], it is natural to speculate that the B-form is adopted for DNA replication whereas the A-form is adopted for transcription. Similar considerations apply in the case of synthetic polynucleotides: a polynucleotide with a highly repetitive base sequence can be found in the S form under conditions in which a natural DNA with essentially random base sequence would be found in the B form. Highly repetitive sequences flanked on both sides by random sequences are known to exist in natural DNAs and so it is possible that under some ionic conditions the repetitive and random sequences would have different conformations. The potential to exploit such structural differences in control processes mediated by specific DNA-protein recognition mechanisms is clear. Although these speculations have not been confirmed, the fact that conformational transitions may be implicated in such fundamental biological processes is powerful justification for extensive study of the stereochemical pathways of these transitions and the factors which promote and control them.

Although X-ray diffraction methods have proved to be extremely successful in the determination of a wide variety of biomolecular structures, including both globular and fibrous proteins and nucleic acids, polysaccharides and macromolecular assemblies, much of the information obtained to date from such studies is essentially static. These macromolecules, like the biological organisms composed from them, are, however, dynamic systems with, for example, modes of vibration, hinge movements and gross conformational flexibility. Although many features of the functions of these biomolecules have in many cases been satisfactorily explained in terms of these static structures, it is clear that a full understanding of biological function depends upon a knowledge of dynamical behaviour and properties. Attempts have been made to determine the areas of greatest dynamical activity within globular proteins for example through the study of anisotropic atomic temperature factors derived from crystallographic data sets collected at various temperatures [26]. However, the most important advance to have been made in this area in recent years has been the widespread availability of intense X-ray beams from electron synchrotrons and storage rings. The high brightness of these machines has reduced, by a factor of one hundred or more, the time required to record a diffraction pattern and this offers the capability to record a series of diffraction patterns from a system while its structure undergoes a transition from one state to another. This clearly represents a major advance in the techniques available to structural molecular biology and has already been exploited in, for example, small-angle solution scattering studies of the assembly of microtubules [27, 28] and chromatin [29], low-angle fiber diffraction studies on muscle contraction [30] and crystallographic studies on the kinetics of protein-substrate interactions using the Laue method [31]. It has also been used to follow conformational transitions in the DNA molecule using high-angle fibre diffraction [32, 33] and it is this work which forms the major topic of this review.

2 Fiber Diffraction Theory and Methods

2.1 Diffraction from Helical Molecules

The Fourier transform of the electron density distribution $\varrho(r, \varphi, z)$ may be written in cylindrical polar co-ordinates in the form:

$$F(R, \Phi, Z) = \sum_{n} \int_{0}^{\infty} \int_{0}^{2\pi} \int_{0}^{c} \varrho(r, \varphi, z)\, J_n(2\pi Rr) \times \exp\left(i\left[n\left(\Phi - \varphi + \frac{\pi}{2}\right) + 2\pi Zz\right]\right) r\, dr\, d\varphi\, dz \qquad (1)$$

where J_n is the n-th order Bessel function of the first kind, (R, Φ, Z) are the cylindrical polar co-ordinates of a point in reciprocal space and c is the length of the repeat along the molecular c-axis [34]. Since a helix is periodic both in the z-direction and in the azimuthal angle φ, its electron density function may be expanded as a two-dimensional Fourier series: —

$$\varrho(r, \varphi, z) = \frac{1}{c} \sum_{n} \sum_{l} g_{nl}(r) \exp i\left(n\varphi - \frac{2\pi lz}{c}\right) \qquad (2)$$

where $g_{nl}(r)$ are the two-dimensional Fourier co-efficients. The Fourier transform of a helical molecule derived from Eqs. (1) and (2) then becomes ([35] and Stokes, unpublished): —

$$F(R, \Phi) = \sum_n G_{nl}(R) \exp in\left(\Phi + \frac{\pi}{2}\right) \tag{3}$$

where: —

$$G_{nl}(R) = \sum_j f_j J_n(2\pi R r_j) \exp i\left(\frac{2\pi\, l z_j}{c} - n\varphi_j\right) \tag{4}$$

In Eq. (4) the summation is taken over all the atoms in the helical repeat unit with co-ordinates (r_j, φ_j, z_j) and atomic scattering factors f_j. The axial periodicity of the molecule confines the Fourier transform to layer-planes in reciprocal space with co-ordinates $Z = l/c$. The combination of axial and azimuthal periodicities imposes a selection rule, given by

$$n = (l - Nm)/K$$

on the Bessel function terms which contribute to the summation in Eq. (3), where N is the number of molecular asymmetric units contained in K turns of the helix and m is any positive or negative integer. All the DNA conformations considered in this paper contain a two-fold rotation axis perpendicular to the helical axis and hence all the terms in Eq. (4) are completely real (assuming that there is no anomolous scattering): —

$$G_{nl}(R) = \sum_j f_j J_n(2\pi\, R r_j) \cos\left(\frac{2\pi\, l z_j}{c} - n\varphi_j\right) \tag{5}$$

When the molecules form a three-dimensional array, the molecular transform is sampled at reciprocal lattice points giving structure factors: —

$$F(h, k, l) = \sum_p \sum_n G_{nl}(R) \exp in\left(\Phi + \frac{\pi}{2} - \varphi_p\right) \times \exp 2\pi\, i(hx_p + ky_p + lz_p) \tag{6}$$

where (hkl) are the Miller indices of a reflection, (x_p, y_p, z_p) are the fractional co-ordinates and φ_p is the azimuthal orientation of the pth molecule in the cell.

Since the crystallites within the fiber are typically in random orientation about the c-axis, the recorded X-ray diffraction has the characteristics of a single crystal rotation pattern and the observed intensities are given by: —

$$I(hkl) = \langle F^*(hkl)\, F(hkl)\rangle_\Phi \tag{7}$$

where the angular brackets denote averaging with respect to the azimuthal angle Φ. Therefore, although the X-ray diffraction from such oriented crystalline fibers is three-dimensional, there is a loss of information arising from accidental and systematic overlapping on a given layer-plane of Bragg reflections which have the same radial co-ordinates. Nevertheless it is possible to obtain detailed three-dimensional structural information from the observed intensity distribution. As a consequence, detailed information on the conformation of nucleic acid double helices and their orientations with respect to the unit cell axes may be derived from crystalline fiber diffraction data.

2.2 Structure Solution and Refinement

There is no direct method for solving the X-ray crystallographic phase problem in the fiber diffraction technique. Hence the main approach to structure solution is to propose a model for the molecular structure, calculate the predicted diffraction pattern and compare it with the observed diffraction pattern. In favourable cases it is straightforward to determine the helical symmetry of the molecule. However, the Bragg reflections in the patterns rarely extend beyond 2.5 Å resolution. Therefore, it is necessary to utilize information on covalent bond lengths and angles derived from single crystal X-ray diffraction studies on small fragments of the nucleic acids in devising the detailed molecular structure at the atomic level. Additional constraints on a stereochemically acceptable model arise from the need to consider non-covalent interactions. There is no guarantee that any such model is unique. Indeed, there have been a number of competing models proposed to account for some DNA diffraction patterns. In particular, there has been a long-standing controversy concerning the D form of DNA with competing models differing from each other in such fundamental features as the handedness of the helix. In these circumstances, it is necessary to have recourse to an objective method of adjudication. Arnott and co-workers [36, 37] have devised an adjudication algorithm, incorporated in the program LALS, which employs the linked-atom least-squares technique to minimize a function of the form: —

$$\Omega = \sum_k e_k \, \Delta\vartheta_k^2 + \sum_m \omega_m \, \Delta F_m^2 + \sum_i k_i \, \Delta d_i^2 + \sum_n \lambda_n H_n \qquad (8)$$

The first summation constrains the conformational torsion angles (and sometimes the covalent bond angles) to lie near to the expected values derived from single crystal and polymer studies; the second summation minimizes the differences between the observed and calculated X-ray structure factor amplitudes; the third summation relaxes unfavourable non-bonded interactions and the fourth summation contains constraints H_n whose values are zero when residue connectivity and furanose ring closure have been achieved.

3 Instrumentation

3.1 The Synchrotron Source

The experiments described in this review were performed at the Synchrotron Radiation source (SRS) at the SERC Daresbury Laboratory. The operating conditions were typically 2GeV electron energy and 200 mA beam current.

The most attractive properties of synchrotron radiation for fibre diffraction experiments are the wavelength tunability, which allows anomolous scattering data to be collected at various wavelengths in the vicinity of an absorption edge, and the high brightness which facilitates the collection of diffraction patterns from a DNA fiber within minutes compared with the several hours required when using a conventional X-ray generator. The latter property is important for the time-resolved fiber diffraction experiments we describe here. Location of ions around the DNA double helix using anomolous scattering in parallel with isomorphous replacement techniques is also in progress in our laboratory.

3.2 The Beamline X-ray Optics

The instrumental requirements for high-angle fiber diffraction are essentially identical to those for protein crystallography. There are two facilities at Daresbury which jointly provide for these techniques: one is located on a dipole bending magnet (line 7) and the other is on the wiggler magnet (line 9). The optical system on line 7 has been extensively described elsewhere [38, 39]. The first optical element on this line is a 0.60 m platinum-coated mirror which is used at a glancing angle of approximately 7.5 mrad giving an observed reflectivity of approximately 40%. The mirror is placed approximately 10 m from the source and provides 1:1 focussing at the camera in the vertical direction.

The second optical element is a triangular Ge (111) monochromator with an oblique cut of 10.5° placed approximately 10 m from the mirror. The monochromator can be bent to provide a vertical demagnification of the source of approximately 10:1 in the horizontal direction at the camera. The wavelength used in the experiments described below was usually 1.488 Å or 1.608 Å. The wavelength spread, $\Delta\lambda/\lambda$, was typically 10^{-3}. The arrangement on the wiggler station is similar.

3.3 The Fibre Diffraction Camera

The beam shape and size were defined by a collimator integral to the high angle fiber diffraction camera. In a typical experiment, a circular beam of 100 μm diameter was used. The camera was placed on an alignment table that was controlled remotely to enable the flux through the camera to be maximized using an ion chamber placed immediately after the collimator. A mylar window containing a backstop at the rear of the camera had suitable dimensions to enable the diffraction pattern to be collected either on film or using a two-dimensional electronic X-ray detector. The camera was flushed with helium which had been bubbled through an appropriate saturated salt solution [40] to control the relative humidity. A service collar contained a probe to monitor the humidity.

3.4 X-ray Detectors

The most convenient X-ray detector is photographic film which has several desirable properties. It has high spatial resolution (~25 μm) and contains effectively an array of fairly efficient detectors — the grains — which act independently to give an extremely high count-rate. X-ray photographic film also has several disadvantages. The process of development blackens some grains which have not been struck by photons giving rise to a background of chemical fog over the entire surface of the film. This background must be removed from the measured intensities. Although this is comparatively straightforward when most of the diffracted intensity is concentrated into strong Bragg reflections, techniques for performing this are more problematic when the diffraction pattern contains significant amounts of diffuse scatter which is often the case when the sample is undergoing a structural transition. A further disadvantage in the context of time-resolved experiments is the mechanical problem of changing the film quickly. In practice this can be achieved in a matter of seconds by placing a set of films on a movable carousel. However, this capability is of limited value since, while a carousel only carries a small number of films (typically eight), a large number of patterns need to be recorded in a time-resolved experiment. The alternative is to change the film manually which in view of the personnel safety requirements generally means a time delay of up to one minute between the end of the collection of one diffraction pattern and the beginning of the next. An additional problem is that the development of a film takes about ten minutes during which time the experiment has continued and the experimenters have been deprived of genuinely real-time information on the current state of the sample. Further, since changes in the diffraction during a structural transition are determined by subtracting successive patterns from each other, it is crucial that there is a very high degree of reproducibility in film location, exposure time and film development conditions. In practice this is difficult to achieve throughout the monitoring of a transition which may involve recording as many as 50 diffraction patterns. Finally, a major disadvantage of film in experiments which require precise intensity measurements for detailed structure determination is its limited dynamic range. This problem is lessened by the use of packs containing several films so that weak reflections are recorded on the first film whereas more intense reflections are attenuated sufficiently by the films in the pack to lie within the dynamic range of one of the under films. However this introduces the need to scale the intensities measured in different films which is a source of systematic error.

Electronic X-ray detectors are currently under development in a number of centers [41]. At Daresbury a multiwire proportional counter (MWPC) with delay-line readout, developed initially at the SERC Rutherford Appleton Laboratory [42], has been used for small-angle scattering, fiber diffraction and protein crystallography. The active area of this xenon-filled detector is 200×200 mm^2. Exposure to intense synchrotron radiation has tended to polymerize the hydrocarbon quencher which then deposits on the anode wires. The wire spacing was originally 1 mm and, although this gave good spatial resolution, the applied voltage was 7 kV and the hydrocarbon deposits gave rise to an unacceptably high trip-rate. The chamber has recently been redesigned and the wire spacing has been increased to 2 mm (Worgan et al., unpublished work). Under these circumstances the chamber can be operated at 2 kV and has now proved to be extremely reliable. However, the loss of spatial resolution has seriously diminished the value of this detector so that it is now insufficient for the measurement of fiber Bragg intensities with a precision adequate for structure determination at the atomic level. This is particularly serious in fiber diffraction analysis where disorder in the fiber has lead to significant arcing and overlapping of neighbouring reflections. However in the brief period when this detector performed to specification it proved to be extremely useful in the collection of time-resolved diffraction data [33]. This was effected by the storage of the data in a fast histogramming memory. The address of each location in memory was given by the concatenation of the spatial co-ordinates (x,y) and the temporal co-ordinate, t. Time-resolution was achieved by incrementing t by one whenever a time period, Δt, had elapsed. In this way it was possible to collect up to 64 time frames each containing 256×256 pixels. In principle Δt may be as small as 1 ms but we have found $\Delta t \sim 60$ s to be suitable for our experiments. A real-time image of the current frame could be displayed on a video monitor which was invaluable in initial studies to determine the behaviour of a sample during a transition. The spatial resolution, although inferior to that of photographic film, proved to be satisfactory in studies of this kind in which highly accurate intensity measurements were not required. An additional advantage of the MWPC is that its dynamic range is limited only by the size of the memory words in which the data are stored. The detector is also effectively noise free and eliminates the need to digitize data which arises when film is used. In our experiments on structural transitions in DNA double-helices we have used both film and the MWPC recording [32, 33].

4 Conformational Polymorphism of DNA

A large number of stable conformations of both natural and synthetic DNA have been observed. They may be characterized in terms of gross structural parameters such as N, the number of molecular asymmetric units in K turns of the helix; h, the axial rise per residue; and r, the axial rotation per residue. Both right- and left-handed helices have been observed [13, 43]. In typical cases the molecular asymmetric unit is a mononucleotide but dinucleotide asymmetric units have been found in molecules in which the chemical repeat consists of two nucleotides [11]. The nucleotide conformations can be related to the different helical parameters both in terms of the backbone and conformational angles and features such as the sugar pucker and the base-pair displacement and orientation with respect to the helix axis.

The A, B and C forms have been observed from fibers of both natural and synthetic DNAs. The A-form [5] is a right-handed helix with eleven nucleotides in one turn of the helix ($h = 2.56$ Å, $r = 32.7°$) (Fig. 1). The asymmetric unit is a mononucleotide. In common with all the conformations described here, the molecule contains a family of diad rotation axes perpendicular to the helix axis which relates one chain of the double helix backbone to the other. Therefore, the two backbones run in opposite directions with respect to the helix axis.

The B-form (Fig. 2) is a particularly well established conformation since it has been observed in a wide variety of crystalline and semi-crystalline forms, in less well-ordered gels, in solution and in whole cells. Within the limits of the accuracy of the data from such specimens, the conformation of naturally occurring DNAs appears to be independent of the precise details of the intermolecular interactions. The B-form is a right-handed ten-fold helix ($N = 10$, $K = 1$, $h = 3.4$ Å, $r = 36°$) with a mononucleotide repeat unit [6]. A number of variants of the B-form of

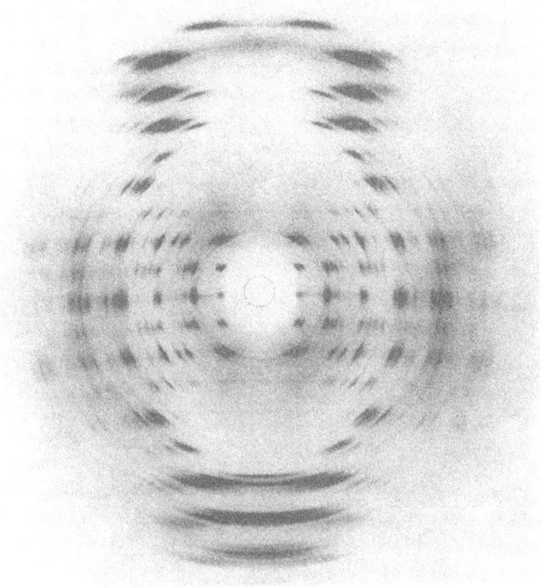

a

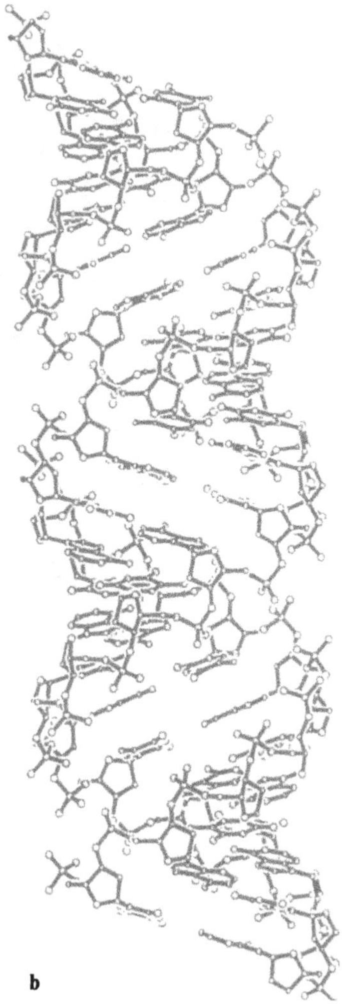

b

Fig. 1a. Diffraction pattern of the A-form from the sodium salt of DNA. **b.** Model of the A-form

synthetic DNAs with regularly repeating base sequences have been reported. In particular it has been proposed that a dinucleotide repeat unit in which the two nucleotides have different conformations giving rise to a "wrinkled" helix accounts for the diffraction patterns obtained from the alternating synthetic polynucleotides poly d(A-T).poly d(A-T) and poly d(G-C).poly d(G-C) [44]. Distinct differences are observed in the B-form patterns obtained from poly d(A-T).poly d(A-T) and poly d(G-C).poly d(G-C) [45]. Model-building studies are currently in progress in our laboratory to define the structural differences in the B-conformations which these two polymers adopt. We have also observed a minor variant, the B″ form, at low humidities with the poly d(G-C).poly d(G-C) molecule [43]. The C-form [7] which is a non-integral right-handed helix is more correctly described as a family of closely related structures distinguished by various values of N and K. More generally the C family appears to be a close structural relative of the B-form.

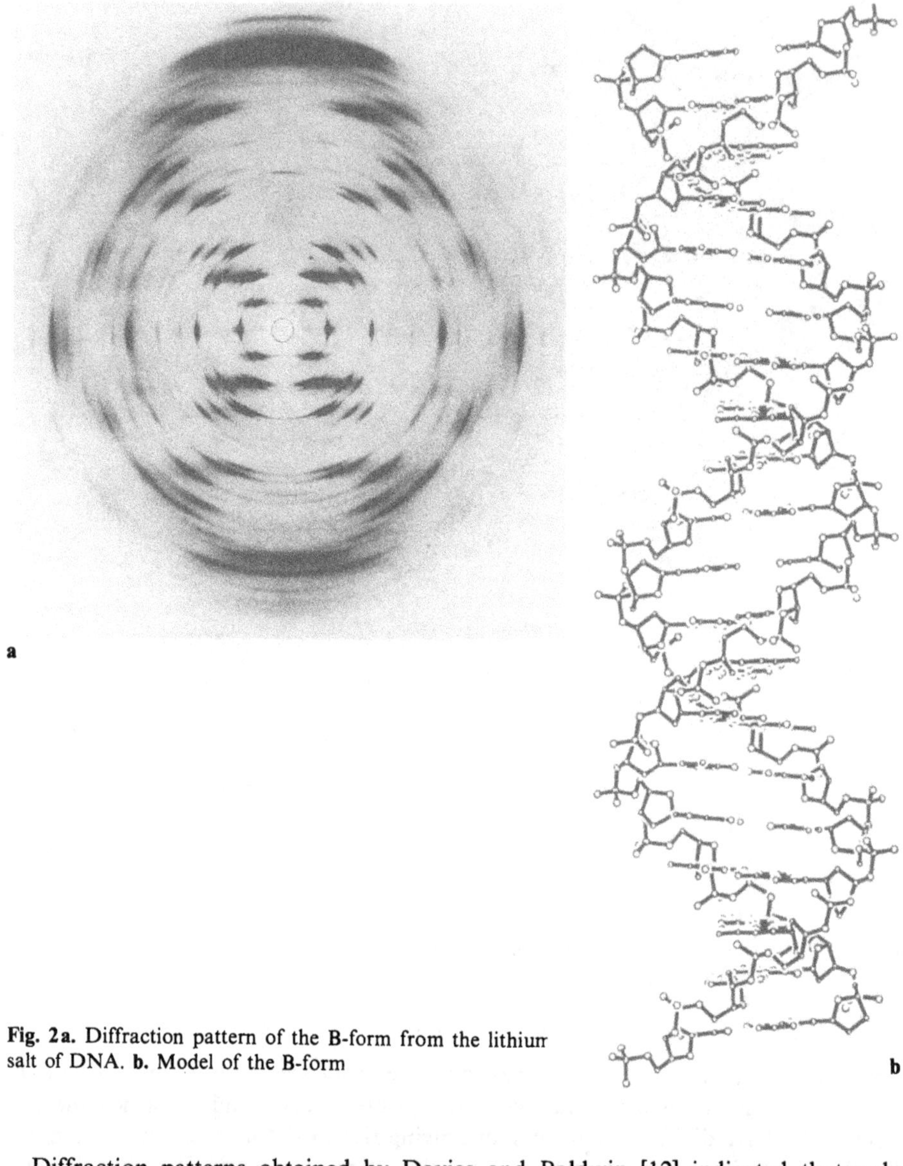

a

b

Fig. 2a. Diffraction pattern of the B-form from the lithium salt of DNA. **b.** Model of the B-form

Diffraction patterns obtained by Davies and Baldwin [12] indicated that poly d(A-T).poly d(A-T) could assume a conformation with axial periodicity about 24.5 Å which they named D-DNA. Similar diffraction patterns were obtained from poly d(I-C).poly d(I-C) by Mitsui et al. [46]. They proposed a novel left-handed double helix with anti-parallel strands and Watson-Crick base-pairs. This eightfold helix contained sugars with an unusual pucker and the bases were placed about 2 Å behind the helix axis i.e. on the opposite side from the position of the base-pairs in the A form of DNA. Arnott et al. [47] obtained better quality D-form diffraction patterns from poly d(A-T).poly d(A-T) which they analysed using the linked atom least squares technique. They rejected the left-handed model of Mitsui et al. [46] and claimed

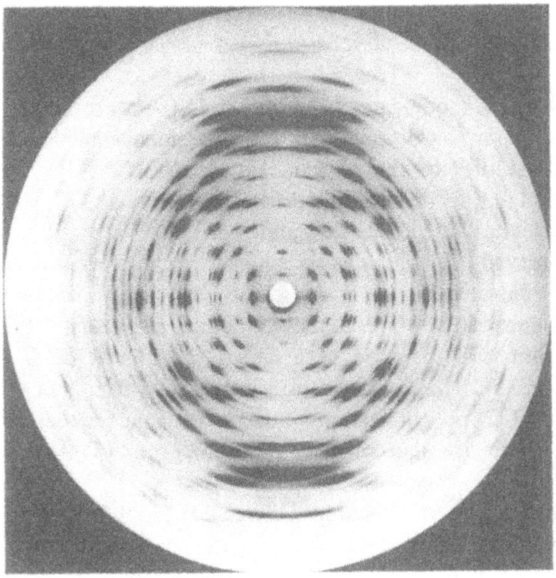

a

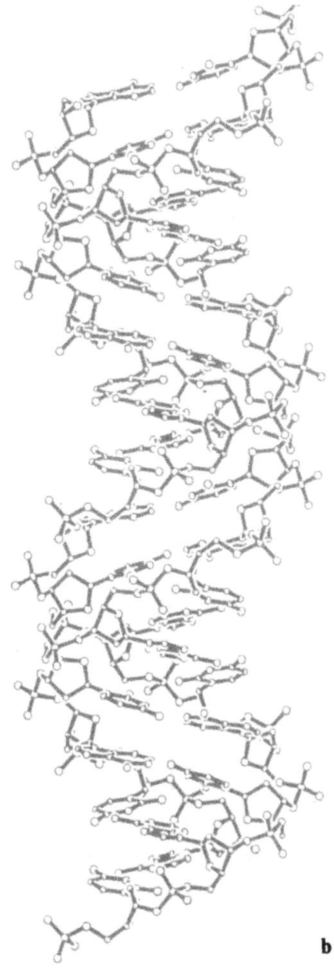

b

Fig. 3a. Diffraction pattern of the D-form from the lithium salt of poly d(A — T) · poly d(A — T) **b.** Model of the D-form

instead that the molecule was a right-handed eightfold helix packed in a tetragonal unit cell with a mononucleotide asymmetric unit (h = 3.0 Å, r = 45°). Refined left-handed models for D-DNA and a sevenfold helical model with Hoogsteen [48] rather than Watson-Crick base-pairs have recently been proposed [49, 50]. We have obtained extremely high quality diffraction patterns from the D-form with Bragg reflections extending to 2 Å resolution (Fig. 3). These patterns have been used with the linked atom least squares technique to compare the agreement with the diffraction data of a range of stereochemically possible models for D-DNA differing in the handedness of the helix, the sugar pucker and the number of nucleotides per asymmetric unit. These refinements and the work to be described in the next section support the contention that D-DNA diffraction patterns can be accounted for by a right-handed helix with 24.0 Å pitch (h = 3.0 Å, r = 45°) in which the repeating unit consists of a dinucleotide. The crystallinity of the diffraction patterns used in this analysis was sufficient to show that the earlier designation of the unit cell as tetragonal was in error and that the lattice parameters were in fact a = b = 17.25 Å, c = 24.2 Å, $\gamma = 94.7°$ i.e. the lattice can be described either as monoclinic with a = b or centred orthorhombic.

The increased interest in left-handed models not only for the D-form but also for the A-, B- and C-forms derived from the discovery of a left-handed fragment of synthetic DNA in single crystal studies of the oligonucleotide $d(GC)_3$ [51]. This so-called Z-form was later observed in fibers containing the polymer poly d(G-C).poly d(G-C) [13]. The fibrous conformation, which is designated S-DNA, is similar in most respects to the structure observed in single crystals. Later work has revealed the existence of two similar S conformations (Fig. 4) designated S_I and S_{II} [52]. In both conformations the molecular asymmetric unit is a dinucleotide and each turn of the helix contains six asymmetric units. In the case of S_I-DNA the helical pitch is 42.8 Å whereas in S_{II}-DNA it is 45.3 Å. The discovery of S_I and S_{II} as variants of the basic S conformation of the poly d(G-C).poly d(G-C) polymer parallels the recognition from single crystal studies that there are also variants of the Z form [53]. These have been designated Z_I and Z_{II}.

The high degree of specificity in the relationship between the conformations accessible to a particular DNA molecule and its nucleotide sequence is well illustrated by the occurrence of the D and S forms. The A, B and C forms have been observed with all naturally occurring DNAs which have been studied and also with both the alternating copolymers poly d(A-T).poly d(A-T) and poly d(G-C).poly d(G-C). However the D form but not the S form has so far only been observed with poly d(A-T).poly d(A-T) and conversely the S form but not the D form has been observed with poly d(G-C).poly d(G-C) but not with poly d(A-T).poly d(A-T).

The role played by cations which neutralize the negatively charged phosphate groups in determining the conformations assumed by DNA was recognized at an early stage in X-ray investigations. It was observed that the A form could be obtained from fibers of naturally occurring DNA in which the cation is any alkali metal except lithium. The lithium ion was unique amongst alkali metal counter-ions in inducing the assumption of the crystalline B-form. However, semi-crystalline B structures were observed when the counter-ion was Na^+, K^+, Rb^+ and Cs^+ as well as Li^+.

Of particular interest for an understanding of the factors which are important in determining the conformation assumed by DNA is the fact that there is a strong interaction between the dependence on base composition and sequence, the ionic environment of the DNA and the degree of hydration of the DNA. This is well illustrated by the sodium salt of natural DNA where the relative humidity at which the transition between the A and B forms is induced depends on the ionic concentration. If the fiber contains barely sufficient sodium to neutralise the phosphate groups then the molecule does not change from the A-form to the B-form until relative humidity is raised above 92%. If however the fiber contains substantially more sodium the B-form is assumed at relative humidities as low as 66%.

The transitions observed in the sodium salt of poly d(A-T).poly d(A-T) are of particular interest [45]. When approximately 0.4 ions per phosphate are present in the fiber, the molecule adopts a semicrystalline C-form at 66% relative humidity. A reversible transition to the crystalline A-form is observed when the relative humidity is increased to 86% but an irreversible change to the D-form occurs at

92%. A further increase to 98% relative humidity induces a reversible transition to the B-form:—

$$\begin{array}{ccccccc} C & \leftrightarrows & A & \rightarrow & D & \leftrightarrows & B \\ 66\% & & 86\% & & 92\% & & 98\% \end{array}$$

The interdependence of ionic environment and the degree of hydration with base composition and sequence is well illustrated by comparing the sequence of conforma-

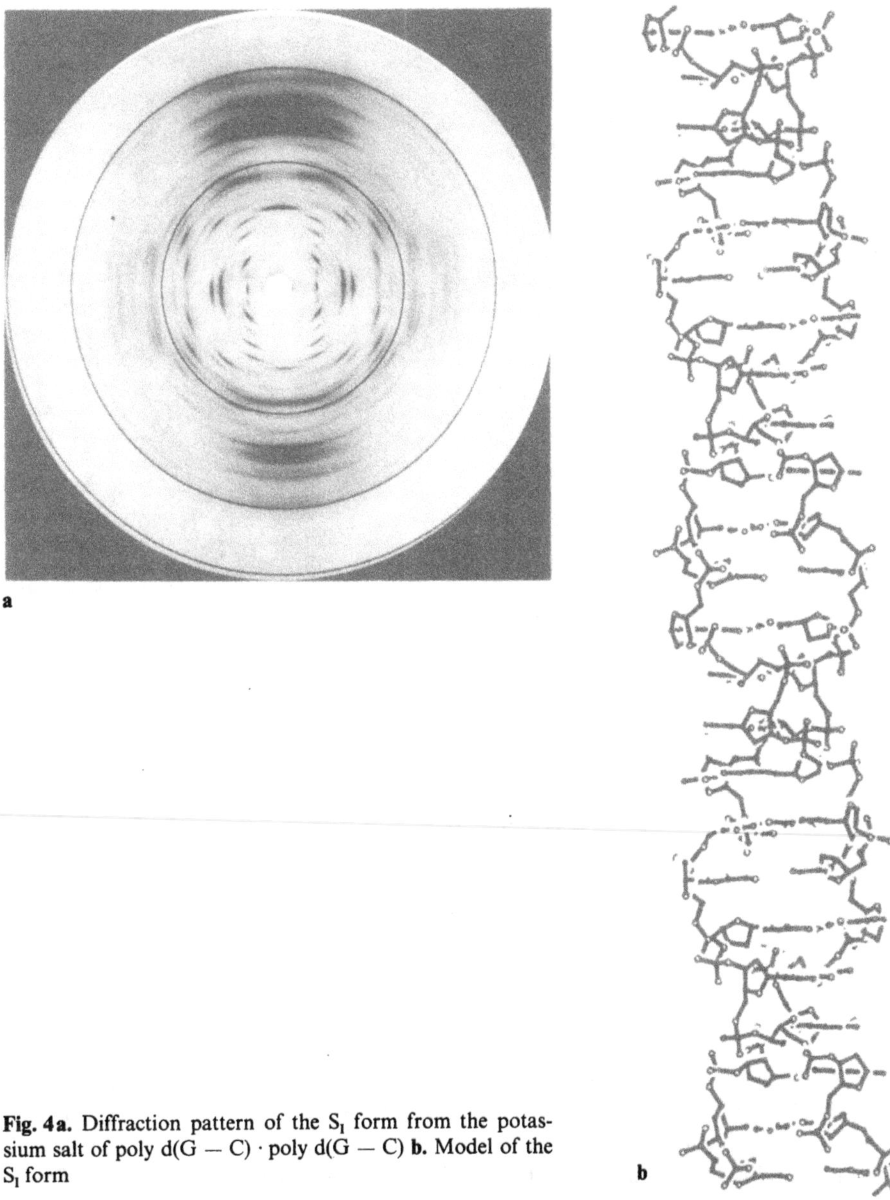

Fig. 4a. Diffraction pattern of the S_I form from the potassium salt of poly d(G — C) · poly d(G — C) **b.** Model of the S_I form

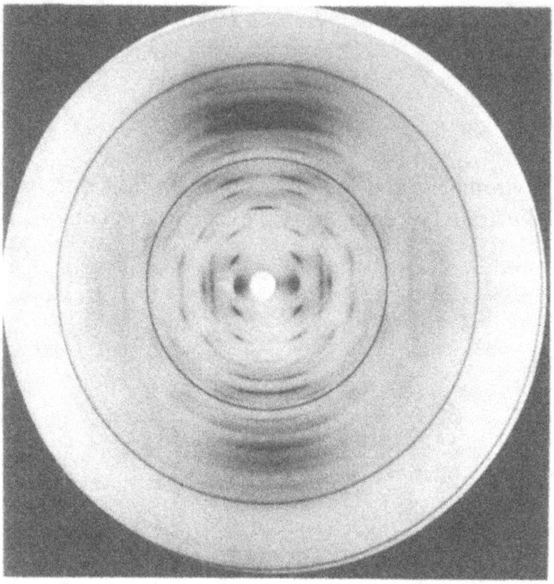

c

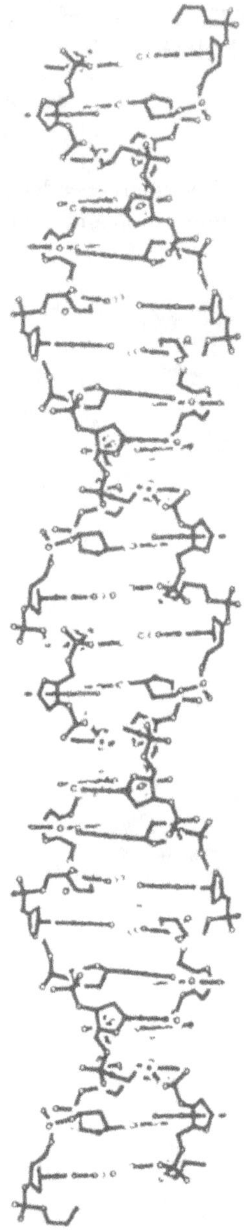

d

Fig. 4c. Diffraction pattern of the S_{II} form from the potassium salt of poly d(G — C) · poly d(G — C) **d.** Model of the S_{II} form

tions observed when the relative humidity of a fiber of poly d(G-C).poly d(G-C) is varied. At 66% relative humidity the conformation of poly d(G-C).poly d(G-C) is B″ with the conformation changing to A at about 75%. This is followed by a transition to the S form at about 92% and then to the B form at 98% relative humidity. The sequence of transitions observed when the relative humidity is reduced depends on the rate at which this occurs [43]. If it is reduced slowly from 98% to 92% then the

S-form is regained but if it is reduced from 98% to 75% without equilibration at 92% then the A-form is regained:—

$$\begin{array}{ccccccc} B'' & \leftrightarrows & A & \rightarrow & S & \rightleftarrows & B \\ 66\% & & 75\% & & 92\% & (\text{slow}) & 98\% \end{array}$$

(B at 98% → A at 75%: fast)

The observation of distinct S_I and S_{II} variants of the S conformation depends on careful control of the relative humidity in the region of 92% to 98%, i.e.

$$\begin{array}{ccccc} S_I & \leftrightarrows & S_{II} & \leftrightarrows & B \\ 92\% & & 95\% & & 98\% \end{array}$$

The influence of base sequence, hydration and ionic environment on the conformation assumed by the DNA double-helix is not unexpected. If the DNA molecule is stabilised in a particular conformation by a particular type and concentration of cation and by a particular concentration of water, then changing the water concentration will perturb the electrostatic interaction between the phosphate groups and could shift the potential energy surface of the system. The observed conformational transitions can be viewed as the response of the system to such potential energy changes. It is clear that the interactions between the DNA, water and ions are complex and that we require detailed structural information not only on the conformation of DNA but also the positions of the ions and water molecules. We are undertaking parallel studies to determine the location of alkali metal ions around the D form of the DNA double-helix using X-ray isomorphous replacement and anomalous scattering methods. In addition we are using neutron isotopic difference techniques based on H_2O/D_2O substitution to locate water molecules. These studies provide static information which complements the dynamic information we have obtained using the time resolved X-ray diffraction experiments on conformational changes which is discussed in the next section.

5 Time Resolved High Angle Fiber Diffraction Studies on Conformational Transitions in the DNA Double Helix

The ability to control conformational transitions within polynucleotide fibers has recently been combined with the high brightness of the SRS in order to record a series of diffraction patterns as the lithium or rubidium salt of a poly d(A-T).poly d(A-T) fiber changed from the B-form to the D-form [32, 33]. Figure 5 shows the time course of the experiment. During a period of 17 hours approximately 100 diffraction patterns, each taking 5 minutes, were recorded on photographic film. In the figure, the shaded regions represent times during which the patterns were recorded whereas the gaps represent times when the experiment was suspended whilst the storage ring was refilled. Four D → B → D transitions were observed with each transition taking approximately one hour during which between 8 and 10 diffraction patterns were

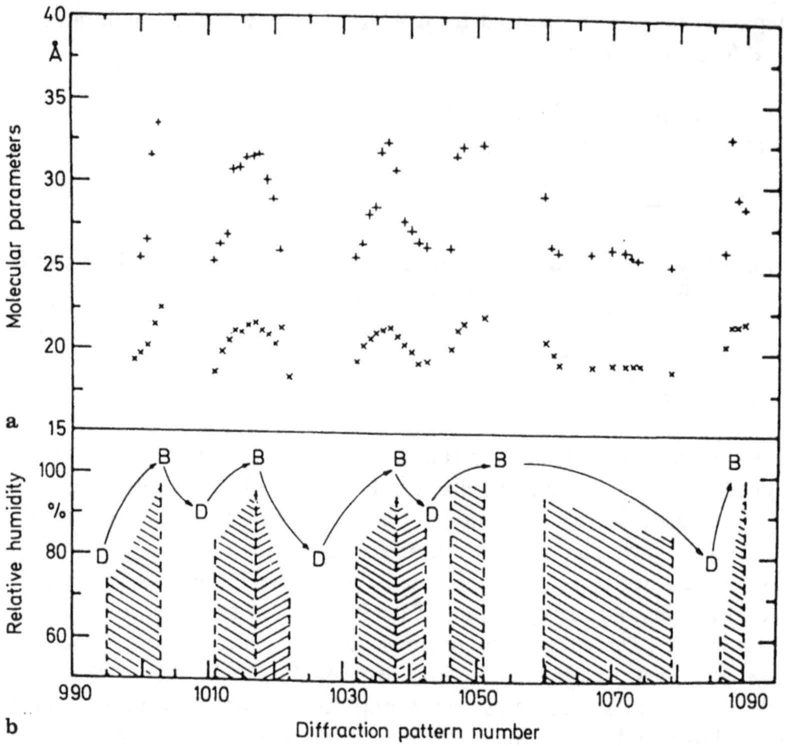

Fig. 5a, b. Schematic diagram of the D → B → D transition experiment. **a.** variation in the lateral intermolecular spacing (×) and the helical pitch (+) of the intermediate semicrystalline form as a function of relative humidity. **b.** Variation of the relative humidity of the fiber environment and molecular conformation during the experiment. The experiment lasted approximately 17 hours during which four reversible transitions between the crystalline D-form and the semicrystalline B-form were observed. In the gaps between the shaded regions the conformation was held constant in either the D or B form

recorded. Although this represents a relatively coarse time-resolution, a complete record of this reversible transition could be compiled by pooling patterns recorded during all four D → B → D sequences. It can be seen from the figure that the rate of the transitions could be controlled by varying the rate of change of relative humidity.

Figure 6 shows a selection of diffraction patterns obtained during D → B transitions whilst the relative humidity was being increased. In the initial state the fibre contained a highly crystalline array of molecules in the D-form (Fig. 6a). Two

---

Fig. 6a–f. Diffraction patterns illustrating typical stages in the D → B transition. In **a.** the reflection marked I is related to the lateral intermolecular separation and that marked II is related to the helical pitch of molecules in the D-form. In **c.** the reflection marked III is related to the lateral intermolecular separation and that marked IV is related to the helical pitch of molecules in the semicrystalline intermediate form. In the reproduction of patterns **b, c,** and **d,** diffraction in the central region of the pattern has been attenuated so that neighbouring reflections may be more easily resolved ▶

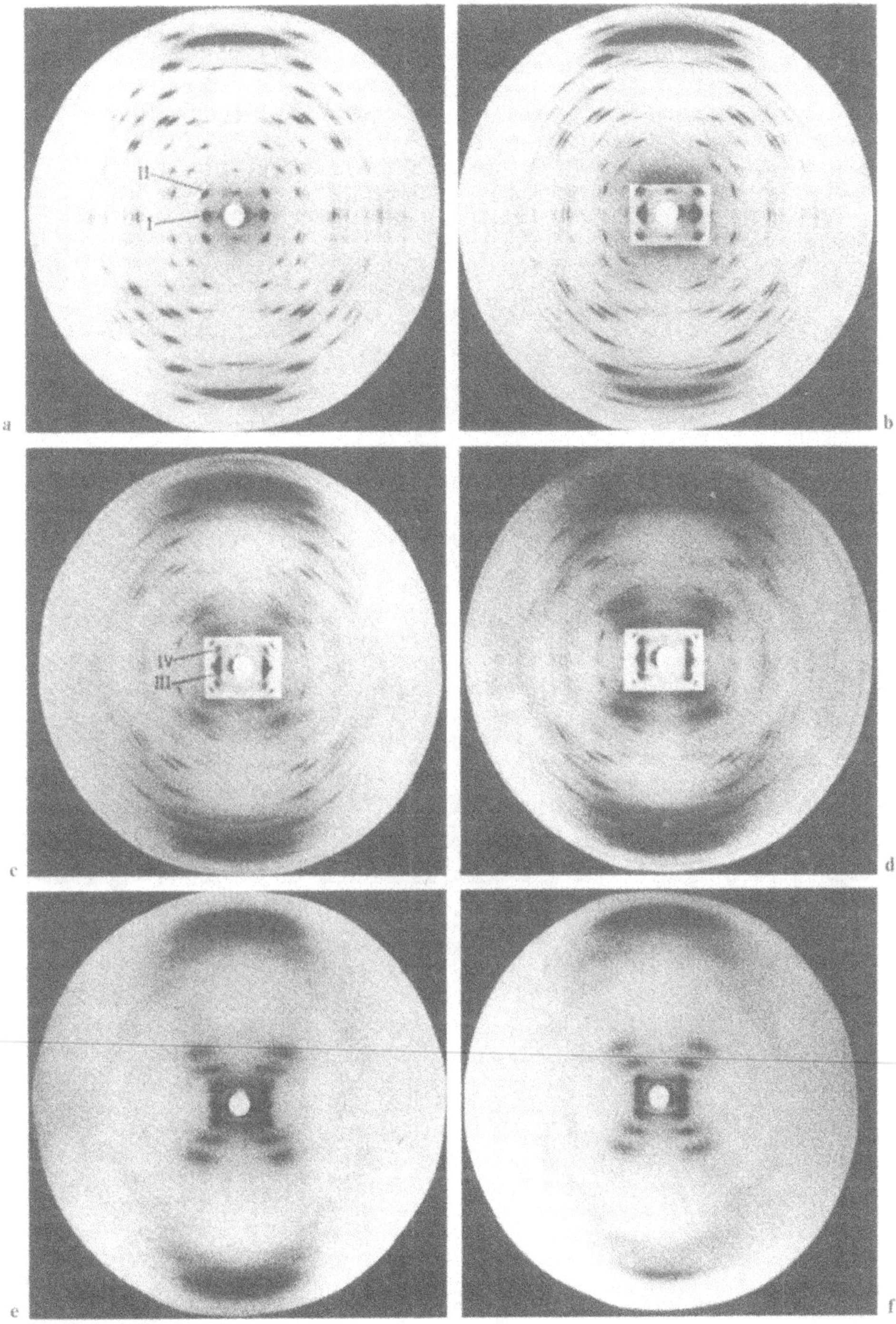
a
II
I
b
c
IV
III
d
e
f

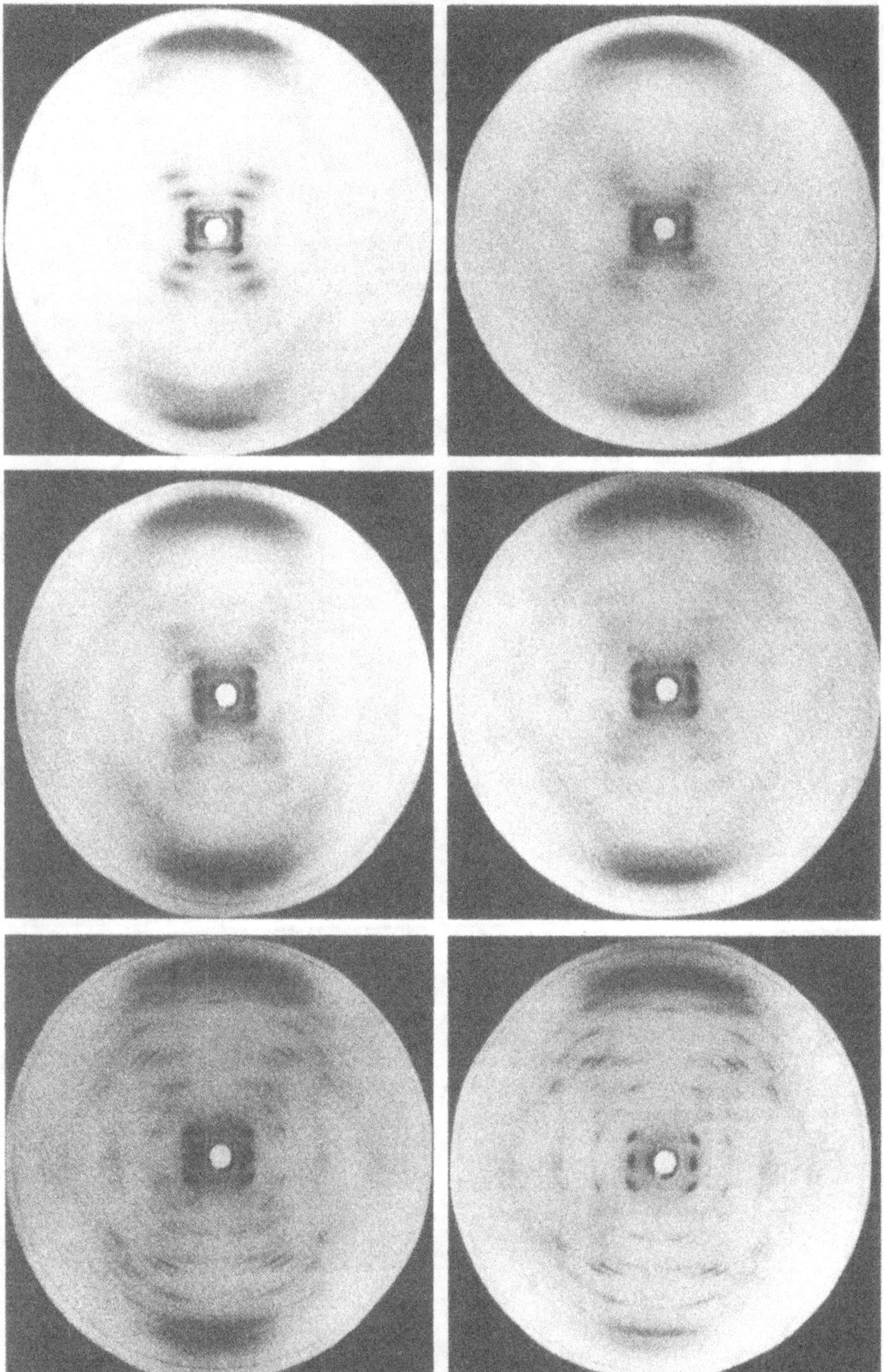

Fig. 7. Diffraction patterns illustrating typical stages in the B → D transition

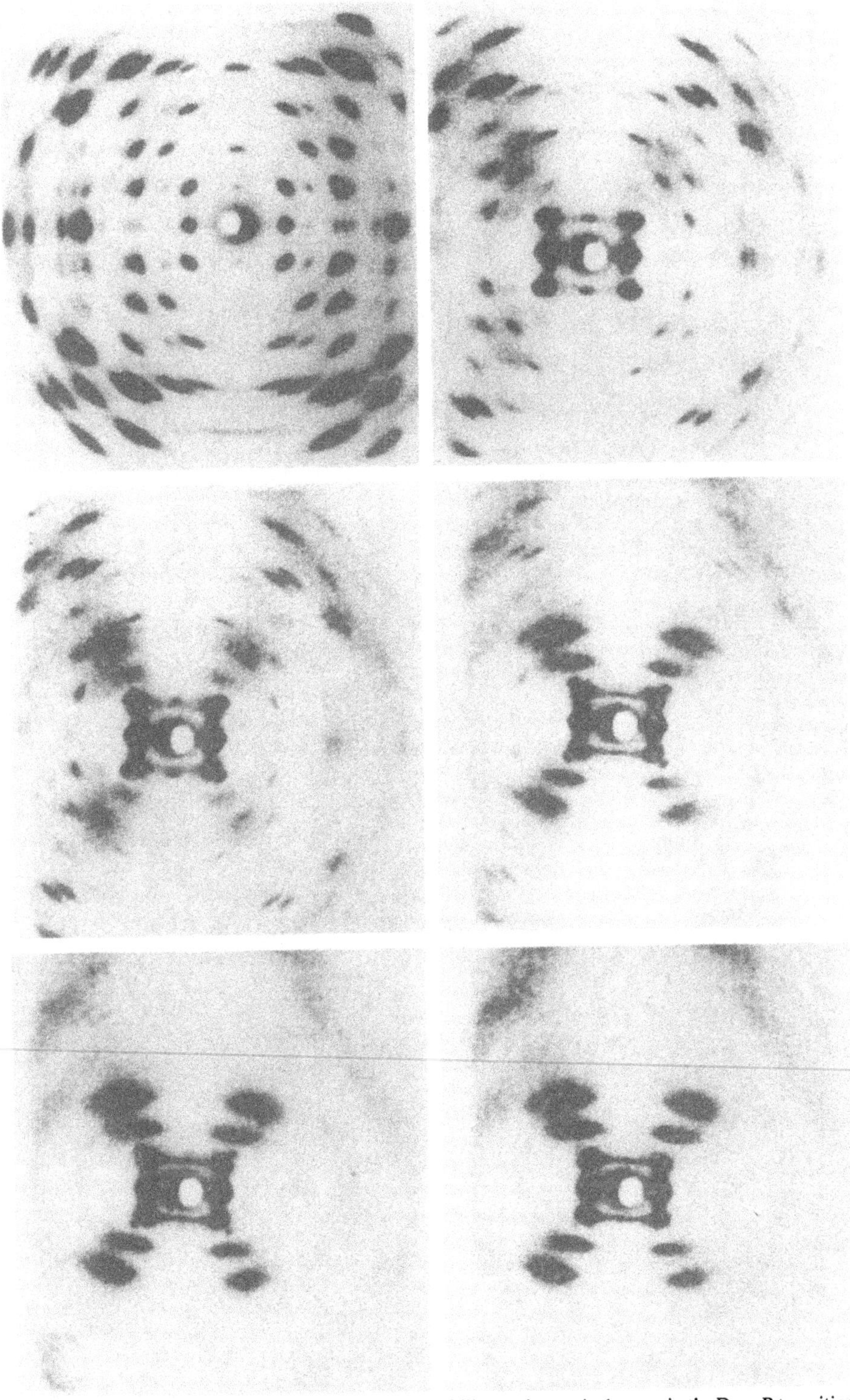

Fig. 8. Diffraction patterns recorded using the MWPC illustrating typical stages in the D → B transition

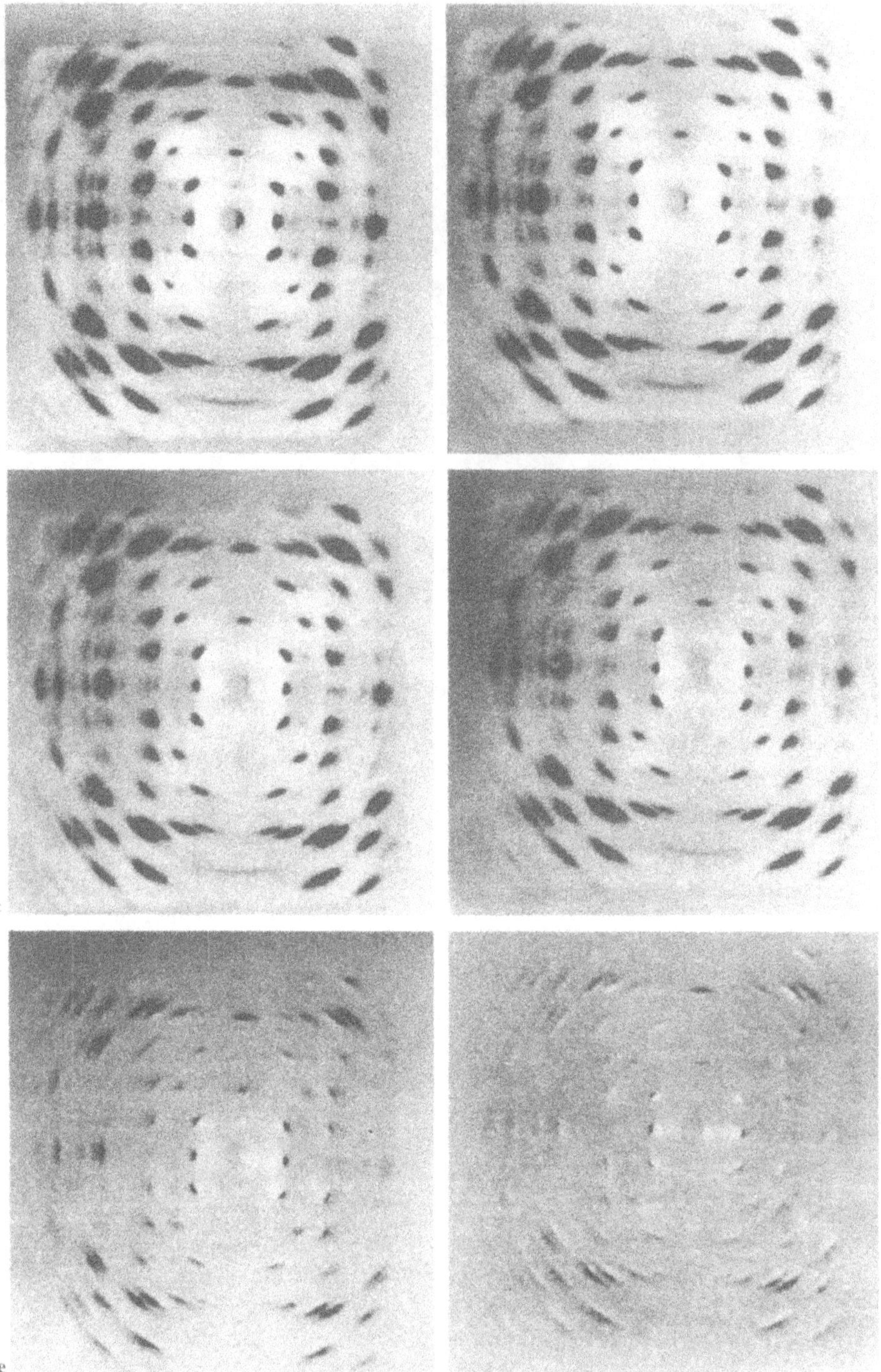

a b c d e f

of the more intense reflections from the D-form have been identified with the labels I and II in this pattern. Reflection I is a measure of the intermolecular separation between adjacent helices whereas reflection II is a measure of the pitch of the helix. The intensities of these two reflections in subsequent patterns in the transition series are proportional to the number of molecules remaining in the D-form at any time. However, the spacings to which these reflections correspond are relatively independent of the relative humidity: the intermolecular spacing is 17.5 ± 0.2 Å and the pitch is 24.3 ± 0.3 Å.

As the relative humidity is increased it is possible to see satellite reflections developing in the vicinity of spots I and II. Two of these have been identified in the diagrams with the labels III and IV. The satellite reflections arise from an intermediate conformation which, in the early stages of the experiment, co-exists with the D-form. Once again, the intensities of these reflections are a measure of the number of molecules in this intermediate conformation. However, it can be seen that as the relative humidity increases the positions and relative intensity of these reflections vary in a smooth and continuous fashion.

The main features of these patterns as the experiment proceeds can be summarised as follows:

1. The intensities of reflections I and II become progressively weaker as the relative humidity increases. However, the positions of these spots do not change. This is consistent with a gradual reduction in the number of molecules in the D-form but this is not accompanied by any change in the gross helical parameters of this conformation.
2. Reflections III and IV become gradually more intense. Reflection III moves equatorially towards the centre of the pattern and reflection IV moves approximately radially towards the center of the pattern. This indicates that the pitch and intermolecular separation of the intermediate conformation gradually change as the number of molecules in this conformation increases.
3. When the relative humidity reaches 98% the layer-line spacing of reflection IV is 33.5 Å which is characteristic of the B-form. Further increases in the relative humidity do not modify the pitch. However, the intermolecular separation continues to increase as the amount of solvent surrounding the molecules rises. In addition to these changes, it is clear that there is a gradual increase in the cross-shaped pattern characteristic of the B-form. Figure S shows that there is a strong correlation between the pitch and intermolecular separation of the intermediate conformations and also that the sequence of events in the D → B transition is highly reproducible. It appears therefore that the DNA molecule does not undergo an immediate transition

◀ **Fig. 9a–f.** Patterns obtained by subtracting the initial D-form diffraction pattern recorded on the MWPC from others obtained during the D → B transition. **a.** The difference between the initial (D-form) pattern and the final (B-form) pattern. The D-form pattern is illustrated clearly by the dark reflections and the semicrystalline B-form cross can be seen in white in the centre of the pattern. **f.** The difference between the initial D-form and a pattern collected almost immediately after illustrates the small change in lattice parameters which occurs at the beginning of the transition. **b, c, d,** and **e** show intermediate stages into the B-form once it is released from the stereochemical constraints imposed by the D-form lattice but rather that an intermediate with a particular pitch is stabilised at any given relative humidity. It is also significant that the values of the pitch observed for these intermediate conformations are not limited to particular discrete values with other values "forbidden". Instead, there is a continuum of pitch values between those characteristic of the B- and D-forms.

Figure 7 shows a compilation of diffraction patterns recorded during B → D transitions which were induced by a reduction of the relative humidity. The behavior of reflections III and IV is similar to that described above for the D → B transitions. However the changes in the diffuse scattering are more complex. Initially (Fig. 7a) the molecule is in a well-defined semicrystalline B-form and, in the centre of the pattern, the diffuse scatter is concentrated on layer-lines 2 and 3. As the relative humidity falls, these layer-lines become less sharp and diffuse scatter appears in the region between them. When the layer-line spacing of reflection IV is between 31 Å and 34 Å (Fig. 7b, c, and d) the diffuse scatter retains the cross-like shape of the B-DNA transform, but when the spacing falls to between 24 Å and 27 Å this scatter is similar in overall intensity distribution to the D-DNA diffraction pattern (Fig. 7e). Sharp reflections begin to appear (Fig. 7e) with spacings characteristic of the D-form and grow in intensity and sharpness until a fully crystalline D form is attained (Fig. 7f).

It is clear that both the B → D and D → B transition proceed via smooth and continuous changes in the polynucleotide conformation. They are quite incompatible with changes in the handedness of the helix during the transition. Two general types of model can be considered in order to justify this assertion. In the first model one envisages the right-handed B form gradually unwinding to form a step-ladder conformation and then rewinding until the left-handed D-DNA conformation is achieved. It should be noted that a transition of this form involves a gradual increase of the molecular pitch from the 34 Å of the B-form, through infinity for a step-ladder conformation to the 24 Å of the D-DNA conformation. The occurrence of such a sequence is clearly inconsistent with the observation that the molecular pitch never falls below 24 Å nor rises above 34 Å during the course of the B → D transition. The second type of model is based on that suggested by Rich and co-workers [51] to account for the transition from the right-handed B-form to the left-handed S-form. In this, hydrogen bonds between base-pairs break in limited regions of the helix so that one of the bases may flip over with a change in orientation about the sugar-base glycosidic link from the *syn* to the *anti* conformation. These regions of transition, which may only be a few base-pairs long, then travel along the helix until the entire molecule is in the D-form. A transition model of this type requires that both B- and D-form molecules should be present simultaneously. The diffraction pattern from such a structure should therefore consist of a mixture of the semicrystalline B and crystalline D forms. This is also inconsistent with our observation of a gradual and continuous change from the B- to D-forms via a set of intermediate conformations. Since neither of these general models can account for a change in helical handedness in a way which is consistent with the diffraction data, it appears that both the B- and D-forms must have the same handedness. As it is generally accepted that the B-conformation is a right-handed helix, these observations demonstrate that the D form is also right-handed and in particular allow the rejection of the left handed models for the D form proposed by Gupta et al. [50] and Drew and Dickerson [49].

The observations described above were based on recording successive high angle X-ray fiber diffraction patterns during the D → B conformational transition in the poly d(T-A).poly d(A-T) double helix using photographic film. While the data recorded in experiments using photographic film have significantly better point to point resolution than data recorded using currently available multi-wire area detectors, such detectors do allow improved time-resolution. This power of such detectors is well illustrated by the selection from a sequence of patterns recorded in a parallel study of the D → B transition in which the data was recorded using a multi-wire area detector (Fig. 8). Figure 9 illustrates difference patterns obtained by subtracting successive frames in Fig. 8. Because of the limitations in dynamic range and reproducibility inherent in film recording the further development of time-resolved studies based on difference techniques is heavily dependent on the availability of reliable area detectors with a high degree of spatial and time resolution.

6 Future Possibilities

In this review we have illustrated the use of synchrotron radiation in time-resolved high angle X-ray fiber diffraction studies of the D ⇆ B structural transition in the poly d(A-T).poly d(A-T) double-helix. The power of this approach is well illustrated by the discovery of an intermediate structure whose helix geometry was shown to vary continuously during the transition. However impressive though this application of synchrotron radiation was, it is important to emphasize that the analysis used only a relatively small fraction of the diffraction data recorded during the transition. It is clear therefore that there is very great potential for further development in this approach and in this final section we would like to identify those features in the molecular system being studied and in the experimental techniques being used which have been important for the success achieved to date and consider which aspects of these features limit future possibilities.

6.1 Features in the Molecular System Being Studied

Clearly what can be achieved in time-resolved high angle fiber diffraction studies depends critically on the nature of the structural transition being studied. From this point of view the D → B transition in the poly d(A-T).poly d(A-T) double-helix was particularly well suited for study because:

(a) The transition is between two well-defined stable conformations both of which are well-characterized by X-ray high angle fiber diffraction analysis.
(b) The transition is readily induced by simply changing a single, easily controlled parameter, i.e. the relative humidity of the fiber environment. In studies we have recently made on the crystallization of the synthetic polymer poly-ether-ether-ketone we have recorded the high angle fiber diffraction while both the temperature and degree of stretching of the specimen were varied [54].

(c) The rate at which the transition occurred could be made to match the time-resolution of the diffraction pattern, i.e. a transition time of about 1 hour could readily be achieved allowing 10 or 12 diffraction patterns to be recorded. In contrast we have encountered much greater difficulty in controlling the A ⇄ B transition in naturally occurring DNAs so that it is sufficiently slow for diffraction patterns from intermediate structures to be recorded. Conversely we have found that the S ⇆ B transition in poly d(G-C).poly d(G-C) requires many hours to occur [55]. While this long transition time does mean that time-resolution is not a problem it does require that the experiment is planned so that all the required data can be recorded during the life time of a single fill of the storage ring.

(d) Whilst it is true that some of the problems associated with achieving adequate time-resolution can be eased by using larger specimens and less well-collimated incident X-ray beams, the problems with preparing specimens which are initially structurally homogeneous increase with specimen size.

6.2 Instrumental Considerations

(a) Clearly the availability of X-ray synchrotron radiation sources with a brightness typically two or three orders of magnitude greater than that of a conventional laboratory X-ray source are crucial to this work although in some cases it may be possible to slow down the structural transition sufficiently for it to be followed to some extent using conventional sources. However, in practice conventional sources are likely to make relatively little impact in time-resolved studies and in fact attention is more likely to continue to focus on the development of sources with higher brightness such as the installation of the high brightness lattice at the Daresbury SRS which has resulted in a gain of a factor of 5 in brightness for a typical high angle fiber diffraction experiment.

(b) All the studies described above have used highly monochromatic radiation. For certain transitions such as the A ⇆ B transition in the DNA double-helix where the transition is clearly very rapid we are considering the possibility of using white beam techniques which parallel the currently extremely exciting applications of the Laue technique in protein crystallography. However, in considering such innovations it is important to recognize the use of white beams in recording high angle fiber diffraction does present formidable problems since it is not simply a question of following changes in the intensity of particular reflections while a transition is taking place. For fiber diffraction patterns, changes can also be expected to occur in the position of reflections. Even more difficulties are presented by changes in the continuous scattering stemming particularly from intermediate structures. Nevertheless there are some problems where, by focussing on a particular region of the diffraction pattern, it will be possible to take advantage of what could be a gain of more than two orders of magnitude in time resolution, e.g. by monitoring the variation in equatorial diffraction during the A ⇆ B transition to investigate changes in molecular diameter and packing.

(c) Developments in detector technology are clearly crucial to possibilities for further progress in time-resolved studies of conformational transitions in fibrous macromolecules. Assuming that acceptable standards of reliability are achieved in electronic area detectors, the main aspects of the specification where one may look for improvement are spatial resolution, counting efficiency and data rate both for an individual pixel and for the detector as a whole. For fiber diffraction patterns like those in Figs. 6 and 7 a spatial resolution of about 1 mm is adequate. While progress in counting efficiency may give useful gains, most important advances can expect to be gained from improvements in maximum data rates. It will not be possible to translate gains in brilliance associated with the ESRF or specimens of greater size if progress is not made in significantly increasing the maximum data rate of detectors. In this context it is worth noting that in our recent work on the S ⇆ B transition in the poly d(G — C) · poly d(G — C) double helix we have used the Enraf-Nonius FAST TV detector [55] and found the spatial resolution to be significantly better than that observed using the multiwire detector. In addition, the TV detector achieved a gain in speed of a factor of about 15 compared with the MWPC and of about 15 compared with film.

6.3 Analytical Techniques

The ability to analyse fully the data obtained in time-resolved experiments will also require major developments in analytical techniques. This is particularly important because much of the information crucial to determining the structural changes occurring during transitions will come from analysis of continuous diffraction. Not only does this data have the most difficulties associated with its measurement but also it requires the development of new methods of analysis.

In conclusion is should be emphasised that time-resolved studies like those described here provide information on stereochemical pathways rather than kinetics. This reflects the obvious point that individual molecules are not free but are in an environment where they interact with other molecules. However, it should be emphasised that, as is well illustrated by "kinetic studies" using for example Laue techniques in protein crystallography where the crystal environment is typically more ordered than in a crystalline fiber, the molecular environment in a fiber is not necessarily any less functionally meaningful than that of many kinetic studies in solution.

Acknowledgements: We are grateful for SERC support (through grants GR/D/08623 and GR/D/59109) which have enabled us to use the SERC Daresbury Synchrotron Radiation Source, to Drs. V. T. Forsyth, A. Mahendrasingam, W. J. Pigram and C. Nave for discussion, to Mrs. H. Moors for help with preparation of this manuscript and to Miss C. R. Oates, Miss K. A. Bellamy and Mr. M. Daniels for help with preparation of the figures.

7 References

1. Watson JD, and Crick FHC (1953) Nature 171: 737
2. Wilkins MHF, Stokes AR, Wilson HR (1953) Nature 171: 737
3. Franklin RE, and Gosling RG (1953) Nature 171: 740
4. Hamilton LD, Barclay RK, Wilkins MHF, Brown GL, Wilson HR, Marvin DA, Ephrussi-Taylor H, Simmons NS, (1959) J Biophys Biochem Cytol S: 397
5. Fuller W, Wilkins MHF, Wilson HR, Hamilton LD (1965) J Mol Biol 12: 60
6. Langridge R, Marvin DA, Seeds WE, Wilson HR, Hooper CW, Wilkins MHF, Hamilton LD (1960) J Mol Biol 2: 38.
7. Marvin DA, Spencer M, Wilkins MHF, Hamilton LD (1961) J Mol Biol 3: 547
8. Cooper PJ, and Hamilton LD (1966) J Mol Biol 16: 562
9. Franklin RE, and Gosling RG (1953) Acta Cryst 6: 673
10. Bram S, and Tougard P (1972) Nature 239: 128
11. Leslie AGW, Arnott S, Chandresekaran R, Ratliff RL, (1980) J Mol Biol 143: 49
12. Davies DR, and Baldwin RL (1963) J Mol Biol 6: 251
13. Arnott S, Chandrasekaran R, Birdsall DL, Leslie AGW, Ratliff RL (1980) Nature 283: 743
14. Arnott S, Hutchinson F, Spencer M, Wilkins MHF, Fuller W, Langridge R (1966) Nature 211: 227
15. Fuller W, Hutchinson F, Spencer M, Wilkins MHF (1967) J Mol Biol 27: 507
16. Arnott S, Wilkins MHF, Fuller W, Langridge R (1967) J Mol Biol 27: 525
17. Arnott S, Wilkins MHF, Fuller W, Langridge R (1967) J Mol Biol 27: 535
18. Arnott S, Wilkins MHF, Venable JH, Langridge R (1967) J Mol Biol 27: 549
19. Milman G, Langridge R, Chamberlain MJ (1967) Proc Nat Acad Sci USA 57: 1804
20. Arnott S, Fuller W, Hodgson A, Prutton I (1968) Nature 220: 561
21. O'Brien EJ, Mac Ewan AW (1970) J Mol Biol 48: 243
22. Arnott S, Hukins DWL, Dover SD, Fuller W, Hodgson A (1973) J Mol Biol 81: 107
23. Chandrasekaran R, Arnott S, Bannerjee A, Smith PJC, Leslie AGW, Puigjaner L (1980) In: French AD, Gardner KH (eds) Fibre Diffraction Methods ACS Symp Series 141
24. Arnott S, Chandrasekaran R, Millane RP, Park HS (1986) J Mol Biol 188: 631
25. Wilkins MHF, and Randall JT (1953) Biochem Biophys Acta 10: 192
26. Artymiuk PJ, Blake CCF, Grace DEP, Oatley SJ, Phillips DC, Sternberg MJE (1979) Nature 280: 563
27. Bordas J, Mandelkow EM, Mandelkow E (1983) J Mol Biol 164: 89
28. Mandelkow E, Mandelkow EM, Bordas J (1983) J Mol Biol 167: 179
29. Bordas J, Perez-Grau L, Koch MHJ, Vega MC, Nave C (1986) Eur Biophys J 13: 157
30. Huxley HE, Simmons RM, Faruqui AR, Kress M, Bordas J, Koch MHJ (1983) J Mol Biol 169: 469
31. Hajdu J, Machin PA, Campbell JW, Greenhough TJ, Clifton IJ, Zurek S, Glover S, Johnson LN, Elder M (1987) Nature 329: 178
32. Mahendrasingam A, Forsyth VT, Hussain R, Greenall RJ, Pigram WJ, Fuller W Science (1986) 233: 133
33. Forsyth VT, Greenall RJ, Hussain R, Mahendrasingam A, Nave C, Pigram WJ, Fuller W (1986) Trans Biochem Soc 14: 553
34. Vainstein BK (1966) Diffraction of x-rays by chain molecules, Elsevier
35. Cochran W, Crick FHC, Vand VE (1952) Acta Cryst 5: 581
36. Arnott S, Wonacott AJ (1966) Polymer 7: 157
37. Smith PJC, Arnott S (1978) Acta Cryst A34: 3
38. Helliwell JR, Greenhough TJ, Carr PD, Rule SA, Moore PR, Thompson AW, Worgan JS (1982) J Phys E 15: 1363
39. Nave C, Helliwell JR, Moore PR, Thompson AW, Worgan JS, Greenall RJ, Miller A, Burley SK, Bradshaw JP, Pigram WJ, Fuller W, Siddons DP, Deutsch M, Tregear RT (1985) J Appl Cryst 18: 396
40. O'Brien FEM (1948) J Sci Inst 25: 73
41. Arndt UW (1986) J Appl Cryst, 19: 145
42. Bateman JE, and Connolly JF (1980) Nucl Instr and Meth 173: 525

43. Mahendrasingam A, Pigram WJ, Fuller W, Brahms J, Vergne J (1983) J Mol Biol 168: 897
44. Arnott S, Chandrasekaran R, Puigjaner LC, Walker JK, Hall IH, Birdsall DL (1983) Nucleic Acids Res 11: 1457
45. Fuller W, Pigram WJ, Mahendrasingam A, Forsyth VT, Nave C, Greenall RJ (1984) Daresbury Laboratory Proceedings, DL/SCI/R22, p 106
46. Mitsui Y, Langridge R, Shortle BE, Cantor CR, Grant RC, Kodama M, Wells RD (1970) Nature 288: 1166
47. Arnott S, Chandrasekaran R, Hukins DWL, Smith PJC, Watts L (1974) J Mol Biol 88: 523
48. Hoogsteen K (1959) Acta Cryst 12: 822
49. Drew HR, Dickerson RE (1982) EMBO Journal 1: 663
50. Gupta G, Bansal M, Sasisekharan V (1980) Int J Biol Macromol 2: 368
51. Wang AHJ, Quigley GJ, Kolpak FJ, Crawford JL, van Boom JH, van der Marel G, Rich A (1979) Nature 282: 680
52. Mahendrasingam A, Pigram WJ, Fuller W, Forsyth VT, Greenall RJ (1984) Daresbury Laboratory Proceedings DL/SCI/R22: 109
53. Wang AHJ, Quigley GJ, Kolpak FJ, van der Marel G, van Boom JH, Rich A (1981) Science 211: 171
54. Oates CR, Mahendrasingam A, Forsyth VT, Greenall RJ, Pigram WJ, Fuller W, Barry M, Oldman R, Blundell DJ (1989) in preparation
55. Mahendrasingam A, Denny RC, Forsyth VT, Greenall RJ, Pigram WJ, Fullter W (1989) in preparation

The Laue Method and its Use in Time-Resolved Crystallography

Keith Moffat[1] and John R. Helliwell[2,3,*]

[1] Section of Biochemistry, Molecular and Cell Biology, Cornell University, Ithaca, NY 14853, U.S.A.
[2] Department of Physics, University of York, Heslington, York Y01 5DD, U.K.
[3] SERC, Daresbury Laboratory, Daresbury, Warrington, Cheshire WA44AD, U.K.

Table of Contents

A review is given of the Laue method and its uses in time-resolved crystallography. The history of the method is covered and various fundamental problems, as outlined by W. H. and W. L. Bragg and R. W. G. Wyckoff, are dealt with. The details of the diffraction geometry, wavelength normalization, and energy deconvolution are given. A survey is then made of the use of the method for time-resolved protein crystallographic analysis. Time resolutions from a few seconds to 10^{-10} seconds are feasible.

* Present address: Department of Chemistry, University of Manchester, Manchester M139PL, UK as well as 3.

1 Introduction

The use of SR (synchrotron radiation), which is intense, collimated, and polychromatic, allows rapid data collection from protein crystals. Because of the significant solvent content, most molecules are active in the crystal and it is possible to induce structural changes that are related to biological function. These changes can then be monitored in the crystal with the SR beam, through changes in the structure amplitudes. Rapid data collection has been found to ameliorate radiation damage effects, especially with intense monochromatized SR at short wavelengths.

Monochromatic methods of data collection rely on rotation of the crystal to satisfy Bragg's Law for many reflections, and to obtain integrated intensities. It is feasible to record complete data sets from a protein crystal on a time scale of 1/2 hour or less. Ultimately the total data collection time is set by the mechanical overheads of rotating the crystal and swapping film cassettes.

In the Laue method a polychromatic beam of x-rays impinges on a stationary crystal. A given reflecting plane extracts from the beam the particular wavelength which allows for constructive interference or reflection to occur. Hence, the data collection time is set only by the exposure time, without any mechanical time overhead. With an unfocussed SR beam, exposure times become of the order of a few seconds for a complete or near complete data set, in favorable cases. With the optimization of the source type and beam line optics, exposure times as short as 10^{-10} seconds for one Laue pattern have been achieved.

2 Historical Perspectives

Several fundamental obstacles apparently argued against consideration of the Laue method for quantitative structure determination. These arguments were

(i) the multiplicity or overlapping orders problem [1, 2].
(ii) the wavelength normalization problem [3].
(iii) in the case of a protein crystal, the radiation sensitivity of the sample in the SR beam [4, 5]; Phillips JC, personal communication).

2.1 The Multiplicity Problem

According to W. L. Bragg [1]:

"X-ray analysis started with the Laue photograph. It is too hard to attach a quantitative significance to the intensity of the spots, which are due to the superposition of diffracted beams of several orders selected from a range of white radiation."

According to R. W. G. Wyckoff [2]:

"In general, reflections to be used in establishing the structure of a crystal should involve wavelengths between the low wavelength limit, λ_{min} and $2\,\lambda_{min}$. Otherwise if $n\lambda > 2\lambda_{min}$ an observed reflection may be partly of one and partly of another order. Only when it is known in some other way that planes of particular types do not give reflections in the first one or more orders can higher values of $n\lambda$ be safely employed in intensity comparisons".

These objections are fundamental in the sense that an energy-sensitive detector with the additional properties of high spatial resolution, high count rate and high absorption efficiency is not available. That is, the contributions of different energy to a given spot cannot be experimentally resolved, in a direct manner.

2.2 The Wavelength Normalization Problem

According to W. L. Bragg [3]:

"The deduction of the crystal structure from the appearance of the Laue photograph is a complicated process, because the intensities of the spots do not depend upon the structure alone. They depend also upon the strength of the components in the continuous range of the original beam to which they are respectively due, and each spot may be composed of several orders superimposed. They are also influenced by the different blackening effect of radiation of different wavelengths and complications arise here owing to the absorption of the x-rays by the silver and bromine in the photographic plate. In spite of these difficulties, the Laue photograph can be made a sound method of analysis, and has, for instance, been used with striking success by Wyckoff. Advantageous features are the ease and certainty with which indices can be assigned to the spots, and the wealth of information represented by a single photograph. Nevertheless, the methods which employ monochromatic radiations are more direct and powerful".

Each reciprocal lattice point is stimulated by some wavelength within the experimental bandwidth. A variety of wavelength-dependent factors affect the measured structure amplitude. These factors include the SR spectral profile, the effect of optical elements, sample scattering efficiencies, absorption of components in the beam and detector response.

2.3 Radiation Damage

The expectation that SR would lead to increased radiation damage [4] has not been found to be a problem. On the contrary with monochromatic SR, radiation damage has been ameliorated. In the case of polychromatic radiation the intensity is strong enough that sample heating becomes significant. Based on early experiences on the NINA synchrotron it was concluded that the protein crystal sample would not withstand the full white beam [5]. However the use of a somewhat limited bandwidth by Moffat et al. [6] gave hope that protein crystals may survive in the full white beam. Pea lectin crystals were used as a test case since these crystals are very radiation insensitive. Several Laue exposures of a single crystal of pea lectin were successfully recorded [7, 8].

3 Laue Diffraction Geometry

For a stationary crystal and white radiation with $\lambda_{max} > \lambda > \lambda_{min}$, the reciprocal-lattice points (RLPs) whose reflections can be recorded lie between the Ewald spheres

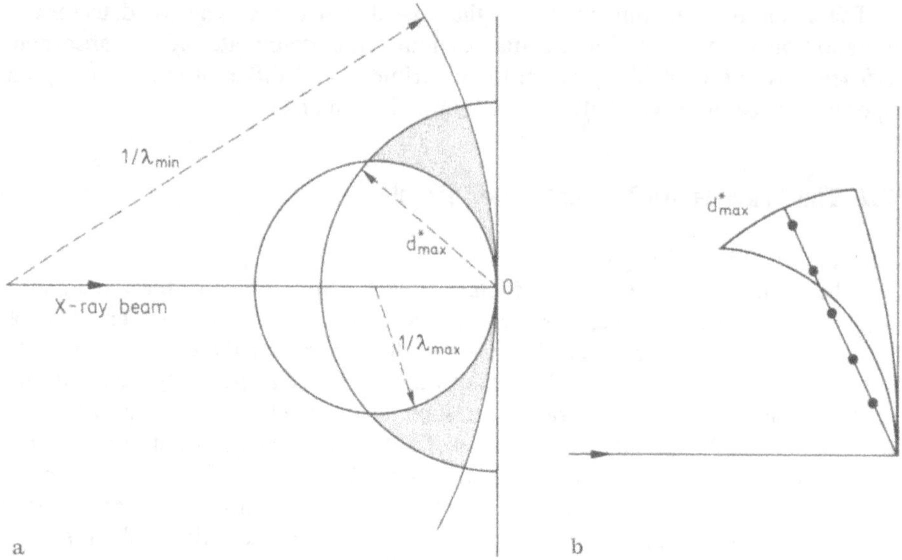

Fig. 1a. Laue diffraction geometry showing the accessible region of reciprocal space between the Ewald spheres associated with λ_{min} and λ_{max} and the sample resolution limited d^*_{max}. O is the origin of of reciprocal space. **b.** A ray with n orders inside the d^*_{min} sphere can have a recorded multiplicity $m < n$ where $(n - m)$ RLPs are outside the accessible region. The diagram shows the case of $n = 5$ and $m = 2$. Only the upper section of the volume of resolution of the accessible region is shown. (From [9])

of radii $1/\lambda_{max}$ and $1/\lambda_{min}$. These spheres touch at the origin of the reciprocal lattice (Fig. 1a), and the wavelength at which any individual RLP diffracts is determined by the reciprocal radius of the Ewald sphere passing through it. There is also a sample resolution limit d^*_{max} ($= 1/d_{min}$), so that no reflections are recorded from RLPs outside a sphere centred at the origin with radius d^*_{max}. The accessible region of reciprocal space, which is cylindrically symmetrical about the incident x-ray beam, may be further limited by experimental restrictions on the scattering angles.

3.1 Energy Overlaps

A Laue diffraction spot can be composed of several RLPs and wavelengths such that for an RLP corresponding to a spacing d and a wavelength λ the RLPs with d/2, d/3, etc. will also be stimulated by the respective wavelengths λ/2, λ/3, etc. That is, these orders of a Bragg reflection are exactly superimposed. This is the effect referred to by Bragg and Wyckoff earlier. Such a spot has a potential multiplicity determined by the resolution limit. However the recorded multiplicity may be less. For example, Fig. 1b shows a ray with five orders inside the d^*_{max} sphere, but of which only two are within the accessible region. The ray and corresponding Laue spot is, therefore, of actual multiplicity 2.

However, contrary to the impression given above by the statements of Bragg and Wyckoff, in general the pattern is dominated by single wavelength, single RLP spots. Only a relatively small fraction of spots arises from multiple wavelengths and multiple RLPs.

If we consider a given reflecting plane of spacing d there is an associated set of Miller indices (h, k, l). Now 2h, 2k, 2l may be beyond the resolution limit and h, k, l may also have no common integer divisor. That is, hkl is an inner point of first order. Hence, the Laue spot that results will contain hkl only.

It can be shown [9] that the probability that a randomly chosen RLP has no common integer divisor is:

$$Q = \left[1 - \frac{1}{2^3}\right]\left[1 - \frac{1}{3^3}\right]\left[1 - \frac{1}{5^3}\right]\left[1 - \frac{1}{7^3}\right]\left[1 - \frac{1}{11^3}\right]\ldots = 0.83191\ldots$$

In the case of $\lambda_{max} = \infty$ and $\lambda_{min} = 0$, all RLPs lying between $d^*_{max}/2$ and the origin of reciprocal space will be recorded as part of a multiple Laue spot. For those RLPs lying between d^*_{max} and $d^*_{max}/2$ there is a probability Q that they will be recorded as singles; this region is 7/8 of the resolution sphere volume.

Hence, a proportion of all RLPs = 7/8Q = 72.8% lie on single rays and a proportion 7/8 = 87.5% of all rays (Laue reflections) are single rays. Similarly, $2(1/2^3 - 1/3^3)\,Q = 14.6\%$ of all RLPs lie on double rays, whereas $(1/2^3 - 1/3^3) = 8.8\%$ of all rays are double rays.

In the case of a more restricted bandwidth, which is the realistic experimental situation, the proportion of singles increases (Fig. 2).

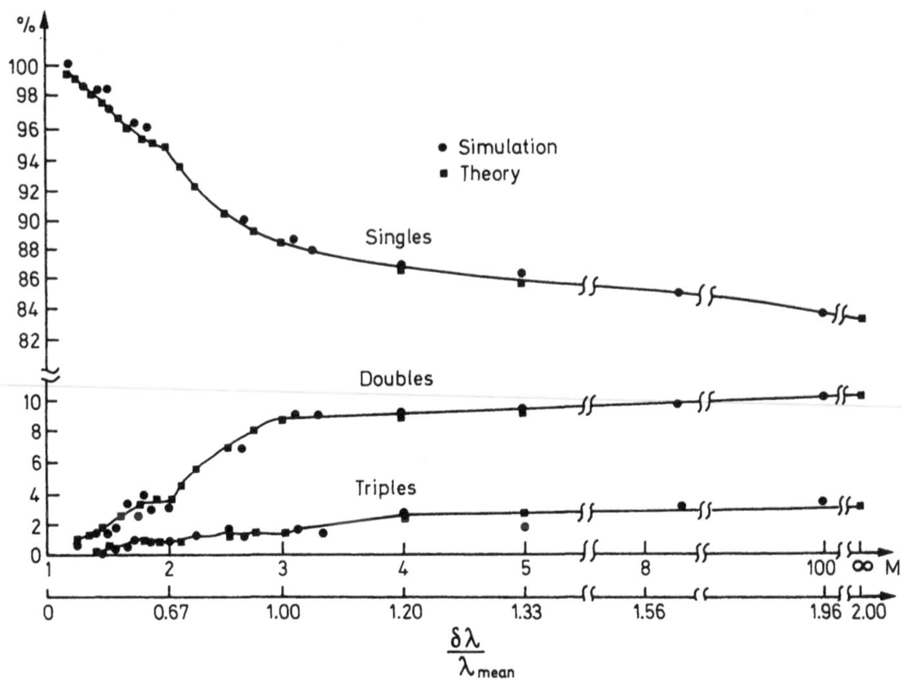

Fig. 2. The variation with M ($= \lambda_{max}/\lambda_{min}$) of the proportions p(1), p(2) and p(3) of RLPs lying on single, double and triple rays for the case of $\lambda_{max} < 2/d^*_{max}$. (From [9])

A probability map can be constructed for the general case (finite λ_{max} and λ_{min}) giving details of the likelihood of RLPs being recorded as single, double, triple, etc. (Fig. 3). A striking feature of the probability map for singles, for example, is the region of zero probability between $d^*_{max}/2$, $2\lambda_{min}$ and the origin.

Moreover, since the volume of reciprocal space sampled is given by:

$$\frac{\pi d^{*4}_{max}\ (\lambda_{max} - \lambda_{min})}{4V^*}$$

where V* = the reciprocal unit cell, the variation of d* to the fourth power shows that the volume of low resolution data sampled is actually quite small.

The Laue method therefore appears very effective at sampling data between d^*_{max} and $d^*_{max}/2$ where RLPs are recorded as singles and need wavelength normalization.

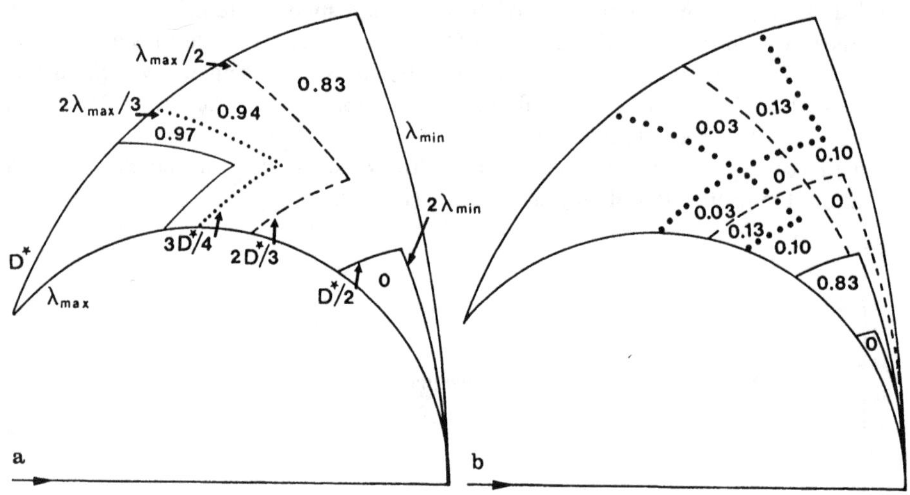

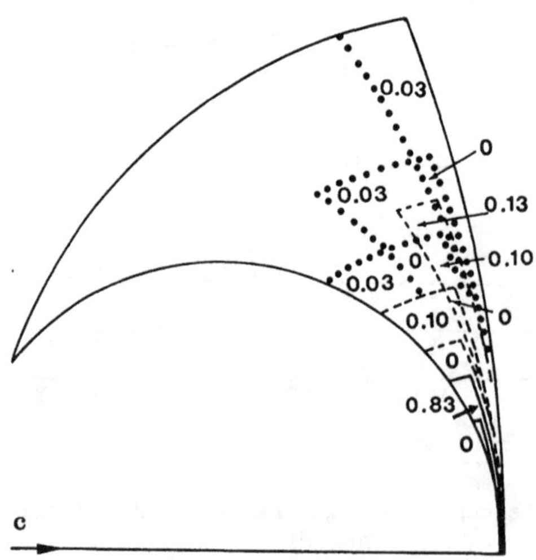

Fig. 3a–c. The probability of a RLP being **a** single; **b** double, and **c** triple. (From [9])

The effective sampling of the region between $d^*_{max}/2$ and $d^*_{max}/3$ needs both energy deconvolution and wavelength normalization.

There are several possible methods available for wavelength (energy) deconvolution. These include:

(a) the use of multiple films to record a Laue pattern. The variation of the film absorption factor with wavelength means that the different components of (and wavelengths in) a Laue spot are attenuated to different extents.

(b) in the case of high symmetry it is likely that a RLP recorded in one case in a "multiple" spot may also occur in a "single" spot. In this case, the other components of a multiple may be measured uniquely. Alternatively, the RLP may occur in several multiples, stimulated by different wavelengths and hence recorded with different intensities. From these, it may be possible to extract the intensities uniquely associated with each RLP.

The first method has been used and reasonable merging R factors obtained although not as good as for singles [10].

In any event, quantitation of Laue photographs is a tractable problem because the pattern is dominated by single wavelength spots.

3.2 Spatial Overlaps

The diffraction spots on a film are of a certain size and if they encroach on their nearest neighbor then these spots may be classed as spatially overlapping. The fraction of spatially overlapping spots actually dominates in the protein crystal case (as compared with the small molecule case). This is especially true in the case of virus Laue patterns (for example, compare Fig. 1 and Fig. 3 of [11]. Computational approaches to the problem involve the use of spatial deconvolution. Experimental approaches involve the use of large detector sizes and long crystal to detector distances; this is feasible because of the tight collimation of the x-ray beam in the situation when the sample is well ordered. This, of course, may well not be the case in a kinetic experiment when the sample may disorder slightly at certain stages, as has in fact been observed [12].

It can be shown (D. W. J. Cruickshank, J. R. Helliwell and K. Moffat, manuscript in preparation) that

(a) the largest average spatial density of spots occurs at

$$\theta_c = \sin^{-1}(0.5\lambda_{min}d^*_{max}) .$$

(b) the largest local, linear spatial density of spots occurs along the arcs approaching nodal spots. (Nodal spots are of potentially low h, k, l value or multiples thereof and are associated with principal zones of the lattice.)

(c) the nearest neighbors to a nodal are single [13]. This explains why the population of spatial overlaps is essentially distinct from the population of energy overlaps (see [8] and Figs. 5c, 5d, and 5e therein).

4 Wavelength Normalization

The measured intensity of a single Laue spot or of an energy deconvoluted spot needs correction or normalization for a variety of wavelength dependent factors if data are to be obtained in a form comparable to or with monochromatic data. This is the case either for pinpointing errors in the data [10] or for using Laue data for structure solving. In the case of monitoring a sample of known structure via the Laue pattern it is possible to use the *fractional* differences whereby the wavelength-dependent corrections cancel. This method is specifically associated with time-resolved measurements and is therefore dealt with below.

The other methods which have been used are:

(a) the use of a silicon standard crystal to derive the wavelength normalization curve or λ-curve [14].

(b) the use of monochromatic data for the same sample as a reference set to derive the λ-curve.

(c) the use of symmetry equivalent reflection intensities measured at separate wavelengths to derive the λ-curve [15].

Methods (b) and (c) have been used to process protein crystal Laue data. The processing statistics are almost as good as monochromatic oscillation data [10, 16, 17].

5 Time-Resolved Macromolecular Crystallography

5.1 Introduction

Transient changes in structure are an integral part of processes such as enzyme catalysis and regulation, cation binding and release, oxygen transport by hemoglobin, and protein folding. Transient structures are typically very short-lived under physiological conditions, certainly seconds or milliseconds or less, and cannot be isolated and crystallized separately. However, they can be generated within a crystal by perturbing an otherwise stable structure, as shown by Parkhurst and Gibson [18] in their pioneering studies of the flash photolysis of crystals of myoglobin and hemoglobin, and by H. W. Wyckoff et al. [19] in their initial flow cell studies of catalysis in single crystals of ribonuclease.

Crystal structures of transient intermediates are accessible to study, provided that substantial concentrations of the desired intermediates can be attained, that their lifetimes are long with respect to the minimum x-ray exposure necessary to obtain a quantifiable diffraction pattern, and that crystallographic isomorphism is maintained throughout the progress of the reaction. Concentrations and lifetimes may be manipulated by adjustment of the chemical nature of the reactants, the pH, ionic strength and nature of the solvent, the presence or absence of activators and inhibitors, by pressure, and of course by temperature. Different intermediates may accumulate to high concentrations, with extended lifetimes, under different experimental conditions; but it may be true that some key intermediates may not accumulate under any conditions.

The Laue technique is aimed squarely at reducing the minimum x-ray exposure, which determines the time resolution in a time-resolved crystallographic experiment. The Laue technique yields the lowest possible exposure time, that is typically two orders of magnitude less than the corresponding monochromatic oscillation exposure time [20]; a large volume of reciprocal space is stimulated simultaneously, which is desirable when the time courses of many thousands of reflections are to be followed; and integrated intensities are obtained from a stationary crystal, without the need for oscillation or precession. All recent practitioners of the Laue technique [6, 8, 12, 14, 21–23] have pointed out that these three advantages fit the Laue technique particularly well for time-resolved crystallography.

The general area of time-resolved diffraction from crystals and less-ordered samples has been well reviewed recently by Gruner [24]. A combination of biochemical and x-ray aspects of time-resolved crystallography is discussed by Hajdu et al. [25] and by Moffat [26], who frovides a more extended version of what follows here.

5.2 The Time-resolved Experiment: Reaction Initiation

The time-resolved experiment contains three key components: reaction initiation, monitoring of reaction progress through changes in x-ray scattering intensities, and data analysis.

Reaction initiation, which may be single-shot or repetitive, must be achieved rapidly and uniformly throughout the crystal of x-ray diffraction size and quality, without damaging the molecules or perturbing the crystal lattice. Speed is necessary to avoid convoluting the true time course with that of the initiating reaction; uniformity avoids initial spatial gradients that would confound later analysis, as would damage; and perturbation of the crystal lattice would at best alter the sampling of the molecular transform and at worst would obliterate the diffraction pattern completely.

In principle, any of the techniques widely used in enzymology to initiate or perturb chemical reactions in solution may be applied to crystals, which in many important respects resemble concentrated solutions. The simplest method of reaction initiation is by diffusion of an essential reactant into a crystal [19]. However, many tens of seconds are required to achieve spatial equilibration across a crystal of typical dimension of 300 μm [27, 28]. Hence, diffusion is only applicable to those few cases where conditions can be arranged so that the subsequent chemical reactions are extremely slow, as in crystals of the highly regulated enzyme, phosphorylase b [12, 25, 29].

For systems that are naturally and reversibly photosensitive such as the carbon monoxide complexes of heme proteins [18, 30, 31], initiation by a light pulse is possible. By preparation of inert but photoactivable reactant or cofactor precursors such as "caged ATP" [32–34], photosensitivity may be conferred on otherwise photoinert systems [21], thus extending the generality of this approach. This approach was first combined with x-ray monitoring by Blasie [35, 36], in studies of oriented multilayers containing the Ca^{2+}-ATPase from sarcoplasmic reticulum.

A third simple and generally applicable method is temperature jump, to perturb a pre-existing chemical or structural equilibrium. However, limitations on thermal conductivity mean that thermal equilibration in response to an external heat pulse is slow [22, 24]. This points to the use of IR lasers that excite the crystal directly (unpublished results of Volz, K., Spirgatis, A. and Moffat, K.).

Satisfying the constraints on reaction initiation is perhaps the most challenging problem in time-resolved crystallography, and it is clear that more work is needed in this area.

5.3 Reaction Monitoring and Data Analysis

If a structural reaction is rapidly and uniformly initiated in a crystal, what will the likely effects be on the Laue x-ray intensities? The very weak intermolecular interactions characteristic of protein crystals make it implausible that existence of a certain tertiary structure in one molecule would favor a particular tertiary structure in its neighbors. That is, all molecules in the crystal are likely to behave independently of each other, as they do in solution. Their populations will evolve smoothly in time in a manner governed by the reactant concentrations and the rate constants. To preserve the x-ray diffraction pattern, spatial coherencc must be maintained between molecules, but temporal coherence need not be. Thus, at any instant the crystal will contain many different conformations, each representing a different structural intermediate. The total structure factor for a particular reflection will be a weighted vector sum of the time-independent structure factors of each conformation, where the weights are the time-dependent fractional occupancies. In favorable cases [26], it may be possible to extract the individual time-independent structure factors from this sum, and hence to obtain directly the structure of each intermediate.

The rms fractional changes in intensity during a time-resolved experiment may be estimated to lie in the 5–20% range, based on comparison of the (static) diffraction pattern of a series of enzyme-substrate complexes, or complexes of hemoglobin with various ligands. These changes in intensity form the raw data and must be measured to a precision of 1–3% of the static intensity. A major simplification of the time-resolved Laue experiment over its static counterpart now appears: only fractional changes in intensities are needed [23]. In the static Laue experiment, the relationship between the measured Laue intensity I_L of the reflection k and the structure amplitude $|F(k, 0)|$ is given by:

$$I_L = c(\lambda, r)\,|F(k, 0)|^2$$

where λ is the wavelength that stimulates the reflection k, r is the direction of the diffracted beam, and $c(\lambda, r)$ is a complicated function of numerous experimental variables such as wavelength, polarization, absorption, and detector sensitivity [22]. Determination of this function is the major additional complexity of the static Laue experiment. Now consider the time-resolved Laue experiment and write [22]:

$$F_L(k, \lambda, t) = \sqrt{I_L} = \sqrt{c(\lambda, r)}\,|F(k, t)|\,.$$

Hence: $\Delta F_L(k, \lambda, t) = F_L(k, \lambda, t) - F_L(k, \lambda, 0) = \sqrt{c(\lambda, r)}\,\{|F(k, t)| - |F(k, 0)|\}$

and $\Delta F_L(k, \lambda, t)/F_L(k, \lambda, 0) = \{|F(k, t)| - |F(k, 0|\}/|F(k, 0)|$
which may be rewritten as:

$$\{|F(k, t)| - |F(k, 0)|\} = |F(k, 0|\,\Delta F_L(k, \lambda, t)/F_L(k, \lambda, 0)\,.$$

Here, the proportionality constant $c(\lambda, r)$ is time-independent and drops out, as first noted by Bilderback et al. [23] and successfully applied by Hajdu et al. [12]. $|F(k, 0)|$ may be quantitated in the Laue experiment itself if $c(\lambda, r)$ has been determined, or it may have been obtained in a prior monochromatic experiment. A time-dependent difference Fourier will utilize $\{F(k, t)| - |F(k, 0)|\}$ as coefficients to yield $\Delta\rho(t)$, the change in electron density as a function of time.

This analysis breaks down if $c(\lambda, r)$ is not strictly constant. For example, the cell dimensions may change during the experiment due to thermal expansion, or the crystal may move very slightly, or the x-ray spectrum may vary due to fluctuations in the source or optics. The likelihood of these experimental artefacts and the magnitude of the errors they would introduce if present must be carefully evaluated in an individual experiment. Indeed, the entire experiment may fail if an order-disorder reaction occurs that leads to transient loss of the x-ray diffraction pattern [37].

How may the diffraction intensities be monitored? The prime requirements for a detector are good time resolution, low noise, wide angular acceptance capable of recording an entire Laue diffraction pattern [9], and high local and global count rate capability. As discussed in detail by Gruner [24], no class of detector at present meets all these requirements. Integrating detectors such as film or the recently introduced storage phosphor/image plates [38, 39, 40] meet all but the time resolution criterion. This may be overcome via a simple streak camera [23] in which the detector is moved in its plane during the exposure. Alternatively, the entire detector may be rapidly replaced by fresh detectors in a sequential manner, so that a time course is built up from 20 or more different detector images [38]. These two approaches are compared by Moffat [26].

5.4 Preliminary Results on Single Crystals

Time-resolved structural studies have so far been conducted on only a few crystals. In pioneering, technically challenging studies, Bartunik [31] used a laser to stimulate reversible photolysis of large single crystals of carboxymyoglobin, multiple repetitions to enhance the signal-to-noise, low temperature to slow the reaction, the Hamburg synchrotron as an intense, monochromatic x-ray source, and a linear position-sensitive detector to monitor the time-dependent intensities of a few adjacent reflections simultaneously. Unfortunately, it was not clear whether the observed changes in diffraction intensities on the ms time scale arose from the tertiary structural changes that are presumed to accompany photolysis of carboxymyoglobin or from artefacts such as laser induced heating of the crystals or mechanical jitter.

Related difficulties were encountered in initial studies of the early stages of protein unfolding in single crystals of lysozyme [22]. Partial unfolding was induced by a single-shot temperature jump and the reaction was monitored using the streak camera to follow numerous time-dependent Laue intensities simultaneously. Indeed, substantial variation in intensities after the temperature jump was seen, on a time scale of 0.2 to 6 s. However, crystal movement or even crystal cracking sometimes followed the temperature jump, and changes in cell dimensions are likely in this type of experiment. Both experiments must be regarded as supplying only a demonstration of principle, rather than definitive results.

The latter have been recently supplied, though so far on a very much slower time scale, by the studies of catalysis in the crystal by phosphorylase b using both monochromatic [29] and Laue [12] x-ray techniques. The former monitored the conversion of heptenitol to heptulose-2-phosphate, and oligosaccharide synthesis involving maltotriose and glucose-1-phosphate, using high resolution monochromatic data collected during the course of these very slow reactions. Although in principle the reaction of a single crystal could be monitored throughout, reactions in separate crystals designed to mimic different stages of completion of the desired single reaction were studied. The time resolution was the time to collect a single monochromatic oscillation data set, and varied between 25 and 150 minutes using bending magnet or wiggler sources at the Daresbury Synchrotron. Spectacularly clean and chemically sensible differences in electron density at the catalytic site were revealed as the reaction progressed (see Fig. 2 of Hajdu et al. [29]). Much superior time resolution was possible via the Laue technique, where individual exposures were only 1 s and a substantially complete data set could be obtained in about 3 s [12]. Unfortunately, the time-resolved picture of the binding of maltoheptose to phosphorylase b could not be obtained due to disorder in the crystals, but an interpretable, static map of the bound complex based on the fractional change in Laue intensities was obtained. These exciting preliminary results leave no doubt that time-resolved Laue studies of catalysis and binding of phosphorylase b are entirely feasible with present-day x-ray sources, on a time scale from 0.5 to 5 s [37].

Very recently, phase transitions in single crystals of the hormone, insulin, have been induced by variation in zinc concentration, and the Laue technique has been used to monitor the change in diffraction intensities at 10 minute intervals over a period of several hours [41].

Time-resolved scattering has been more widely studied in less-ordered systems than single crystals. In an early, important study using oriented membranes, Blasie et al. [35, 36] examined x-ray diffraction from multilayers of the sarcoplasmic reticulum membrane containing the Ca^{2+}-ATPase, and its time-dependence on a 100–500 ms time scale after rapid photolysis of caged ATP released authentic ATP. This appears to be the first example of photoenzymology which used x-ray diffraction to monitor the reaction course. Also, time-resolved x-ray scattering studies of phase transitions induced by changes in hydration of fibers of DNA have been conducted with notable success by Fuller and collaborators [42].

5.5 Future Prospects

Laue experiments will be greatly aided when they are performed on beam lines that utilize the next generation of x-ray wigglers and undulators at the ESRF and the APS, or on prototypes at CHESS and PEP. These beam lines will deliver intensities to the crystal that are two to three orders of magnitude higher than those presently available. That is, exposures now measured in ms will approach μs. More importantly, they will require the x-rays emitted by only a few bunches of positrons (electrons), perhaps only one. Indeed, it has very recently proved possible (43, and unpublished results of K. Moffat and collaborators) to obtain x-ray diffraction patterns using the x-rays emitted by a single bunch of electrons during their traversal of an undulator installed at CHESS, with an exposure time of only 100 ps. Experiments on macromolecular crystals that explicitly utilize the pulsed nature of these sources will be possible, similar to those already conducted on the melting of silicon crystals on the ns time scale [44]. Stroboscopic experiments [23] are limited in time resolution only by the x-ray pulse length, roughly 100 ps, yet may be conducted with integrating detectors such as storage phosphors. The x-ray aspects of such experiments are feasible today; the challenge lies in identifying suitable systems and in devising suitable techniques for reaction initiation and (if necessary) repetition.

6 References

1. Bragg WL (1975). In: Phillips DC, Lipson H (eds) The development of x-ray analysis, p 137
2. Wyckoff RWG (1924) The structure of crystals, Chemical Catalog Co, p 142
3. Bragg WL (1949), General survey. In: Bragg WH, WL The crystalline state p 27
4. Blundell TL, Johnson LN (1976), Protein crystallography, Academic Press
5. Greenhough TJ, Helliwell JR (1983) Prog Biophys Mol Biol 41: 67
6. Moffat K, Szebenyi D, Bilderback DH (1984) Science 223: 1423
7. Helliwell JR (1984) Reports on Progress in Physics, 47: 1403
8. Helliwell JR (1985) J Mol Struct 130: 63
9. Cruickshank DWJ, Helliwell JR, Moffat K (1987) Acta Cryst A43: 656
10. Helliwell JR, Habash J, Cruickshank DWJ, Harding MM, Greenhough TJ, Campbell JW, Clifton IJ, Elder M, Machin PA, Papiz MZ, Zurek S (1989) J Appl Cryst, in press
11. Bilderback D, Moffat K, Owen J, Rubin B, Schildkamp W, Szebenyi D, Temple BS, Volz K, Whiting B (1988) Nucl Instrum and Meth A266: 636
12. Hajdu D, Machin PA, Campbell JW, Greenhough TJ, Clifton IJ, Zurek S, Gover S, Johnson LN, Elder M (1987) Nature 329: 178
13. Jeffery, JW (1958) Z Kristallogr 110: 321
14. Wood IG, Thompson P, Mathewman JC (1983) Acta Cryst B39: 543
15. Campbell JW, Habash J, Helliwell JR, Moffat K (1986) Information Quarterly for Protein Crystallography 18, Daresbury Laboratory
16. Machin P (1987) In: Helliwell JR, Machin PA, Papiz MZ (eds) Computational aspects of protein crystal data analysis. Proceedings of a Daresbury Study Weekend DL/SCI/R25
17. Temple B, Moffat K (1987) In: Helliwell JR, Machin PA, Papiz MZ (eds) Computational aspects of protein crystal data anylsis Proceedings of a Daresbury Study Weekend DL/SCI/R25
18. Parkhurst LJ, Gibson QH (1967) J Biol Chem 242: 5762
19. Wyckoff HW, Doscher M, Tsernoglou D, Inagami T, Johnson LN, Hardman K, Allewell NM, Kelley DM, Richards FM (1967) J Mol Biol 127: 563
20. Moffat K, Cruickshank D, Helliwell JR (1987) In: Bianconi A, Congiu Castellano A (eds) Biophysics and synchrotron radiation Vol. 2. Springer Verlag p. 47
21. Moffat K, Bilderback D, Schildkamp W, Szebenyi D, Loane R (1986) In: Chance B, Bartunik HD New methods in x-ray absorption, scattering and diffraction. Academic Press, New York, p 125
22. Moffat K, Bilderback D, Schildkamp W, Volz K (1986) Nucl Instrum and Meth A246: 627
23. Bilderback DH, Moffat K, Szebenyi D (1984) Nucl Instrum and Meth 222: 245
24. Gruner S (1987) Science 238: 305
25. Hajdu J, Acharya KR, Stuart DI, Barford D, Johnson LN (1988) TIBS 13: 104
26. Moffat K (1989) Ann Rev Biophys and Biophys Chem 18: 309
27. Westbrook EM, Sigler PB (1984) J Biol Chem 259: 9090
28. Farber GK, Machin PA, Almo S, Petsko GA, Hajdu J (1988) Proc Natl Acad Sci USA 85: 112
29. Hajdu D, Acharya K, Stuart DI, McLaughlin PJ, Barford D, Oikonomaki NG, Klein H, Johnson LN (1987) EMBO J 6: 539
30. Austin RH, Beeson KW, Eisenstein L, Frauenfelder H, Gunsalus IC (1975) Biochemistry 14: 5355
31. Bartunik HD (1983) Nucl Instrum and Meth 208: 523
32. Kaplan JH, Forbush B, Hoffman JF (1978) Biochemistry 17: 1929
33. McCray JA, Herbette L, Kohara T, Trentham DR (1980) Proc Natl Acad Sci USA 77: 7237

34. McCray JA, Trentham DR (1989) Ann Rev Biophys and Biophys Chem 18: 239
35. Blasie JK, Herbette L, Pierce D, Pascolini D, Scarpa A, Fleischer S (1982) Ann New York Acad Sci 402: 478
36. Blasie JK, Herbette K, Pascolini D, Skita V, Pierce D, Scarpa Á (1985) Biophys J 48: 9
37. Hajdu J, Greenhough TJ, Clifton IJ, Campbell JW, Shrive AK, Harrison SC, Liddington R (1988) Brookhaven Symp Quant Biol, in press
38. Amemiya Y, Wakabayashi K, Tanaka H, Ueno Y, Miyahara J (1987) Science 237: 164
39. Whiting BR, Owen JF, Rubin BH (1988) Nucl Instrum and Meth, A266: 628
40. Bilderback D, Moffat K, Owen J, Rubin B, Schildkamp W, Szebenyi D, Temple BS, Volz K, Whiting B (1988) Nucl Instrum and Meth A266: 636
41. Reynolds CD, Stowell B, Joshi KK, Harding MM, Maginn SJ (1988) Acta Cryst B44: 512
42. Mahendrasingam A, Forsyth VT, Hussain R, Greenall RJ, Pigram WJ, Fuller W (1986) Science 233: 195
43. Szebenyi DME, Bilderback D, LeGrand A, Moffat K, Schildkamp W, Teng T-Y (1989) Trans Amer Cryst Assoc, in press
44. Larson BC, White CW, Noggle TS, Barhorst JR, Mills DM (1983) Appl Phys Lett 42: 282

Structural, Mesomorphic and Time-Resolved Studies of Biological Liquid Crystals and Lipid Membranes Using Synchrotron X-Radiation*

Martin Caffrey

Department of Chemistry, The Ohio State University, 120 West 18th Avenue, Columbus, Ohio 43210, U.S.A.

Table of Contents

The prodigeous X-ray flux and tuneability of synchrotron radiation has proved an enormous boon in studies of the physicochemical properties of biological liquid crystals. Over the past several years synchrotron radiation has been used to advantage in investigations of the dynamics and mechanism of biological liquid crystal (lipid) phase transitions in bulk systems, lyotrope transport between and within bulk mesophases, X-radiation damage of lipid, lipid-lipid and lipid-lyotrope miscibility and of heavy atom location in Langmuir-Blodgett lipid films. A review of these disparate but related investigations is presented in this chapter.

General Introduction

The prodigious X-ray flux and tuneability of synchrotron radiation has proved an enormous boon in studies of the physicochemical properties of biological liquid crystals. Over the past several years synchrotron radiation has been used to advantage in investigations of the dynamics and mechanism of biological liquid crystal (lipid) phase transitions in bulk systems, lyotrope transport between and within bulk mesophases, X-radiation damage of lipid, lipid-lipid and lipid-lyotrope miscibility and of heavy atom location in Langmuir-Blodgett lipid films.

A review of these disparate but related investigations is presented beginning with a description of the use of time-resolved X-ray diffraction (TRXRD) to study lipid phase transition kinetics and mechanism in Sect. 1. It is the enormous intrinsic intensity of synchrotron radiation that enables TRXRD measurements to be made. However, this advantage brings with it the hazards of radiation damage. This critical issue is addressed in Sect. 2 along with recommendations for minimizing the effect.

Before carrying out kinetic measurements it is imperative that the equilibrium properties of the system under study be known. This information is most conveniently presented in the form of temperature-composition (T–C) phase diagrams. Two new approaches, utilizing TRXRD, which expedite T–C phase diagram construction have been developed and are reviewed in Sect. 3.

In the course of acquiring data for constructing T–C phase diagrams by one of the new methods mentioned in the preceding paragraph, the movement of boundaries separating mesophases in a lyotrope gradient can be followed directly by TRXRD. In Sect. 4, the use of this data to determine the mutual diffusion coefficient of water in contiguous hydrated mesophases is reviewed.

Finally, both the tuneability and high photon flux features of synchrotron radiation are taken advantage of in establishing the location, with subangstrom resolution, of heavy atoms in Langmuir-Blodgett lipid films. Here, X-ray standing waves established under both specular and Bragg reflection conditions, are used in the determination. In Sect. 5 an outline of the standing wave method and results to date and an indication as to the future of the method in studies of membrane structure and of the membrane/aqueous interfacial composition is presented.

1 Studies of the Kinetics of Lipid Phase Transitions Using Time-Resolved X-Ray Diffraction

There are many vital cellular processes, such as membrane fusion, which must involve, if only locally and transiently, changes in lipid phase state. To date, however, little attention has been paid to the dynamics of lipid phase interconversions. If, indeed, such changes are physiologically relevant, they must occur on a time scale comparable with those taking place in vivo. Thus, there is a need to establish the kinetics of lipid phase transitions. In addition to the kinetics, little is known about the mechanism of lipid phase transitions. Such information is integral to our understanding of the structural and compositional requirements for transitions that occur in biological, reconstituted and formulated systems.

A method, referred to as time-resolved X-ray diffraction (TRXRD), has recently been developed which facilitates the direct and quantitative measurement of bulk lipid phase transitions. Mechanistic insights are provided by the ability to detect transition intermediates. The method combines the advantages of a focused, monochromatic synchrotron derived X-ray beam and a suitable live-time X-ray imaging device to continuously monitor diffracted X-rays from the transforming lipid. Kinetic and mechanistic studies are performed by using TRXRD in conjuction with relaxation measurements following the imposition of a rapid perturbation in one or more thermodynamic variables. The ultimate goal is to understand the mechanism of lipid phase transitions and the structural inter-relationships of the different mesomorphs. This should greatly augment our appreciation of the physiological role of lipids and our understanding of the physicochemical properties of these liquid crystalline materials.

The method of TRXRD represents what amounts to a technological conjugation of diverse technologies in a way that is extending our knowledge and understanding of material properties. The relevant technologies encompass synchrotron radiation [7, 12, 14, 47, 67, 127, 128, 153], X-ray optics [9, and references therein], sample manipulation and applied variable measurement [15–19, 23, 26, 28, 63, 89, 140, 142], X-ray imaging and data analysis [4, 6, 8, 16, 18, 21, 58, 59, 63, 65, 82, 152]. Each is a discipline unto itself and, as indicated in the preceding references, each has been reviewed previously either singly or in combination with others. The perspective I present here is that of one who has worked in the field since its inception and who recognizes that the method is still in development. Thus, while each experiment provides new information, it is just as likely to reveal new problems and limitations. As a consequence, I emphasize the shortcomings of the TRXRD approach to the study of lipid phase transition kinetics so as to focus attention on the immediate problems and hopefully, to expedite their solution. I begin with a description of the factors motivating the study of lipid phase transition kinetics and an overview of lipid mesomorphism. The principle of and information derived from TRXRD is introduced next along with an outline of certain experimental aspects of these live-time measurements. The limitations of TRXRD will be addressed from four perspectives: the synchrotron facility, the sample, instrumentation and experimental design. This is followed by a summary of results to date and a view to the future.

In preparation for this review a literature search was conducted in June 1988 of the Chemical Abstracts CA file from 1967 to early 1988. The following search logic was used: [(lipid or phospholipid or fat or liquid crystal) and (kinetic or dynamic or rate) and (transition or transform or phase change)]. The search produced 431 references of which 87 were considered relevant.

1.1 Motivation

Despite the fact that life is dynamic, surprisingly little attention has been devoted to the study of the dynamics of vital processes at the level of macromolecular structure and assembly. Fortunately, the situation is changing. In this context, interest has recently been focused on the cell membrane, itself an extremely lively structure. Consider for a moment the structural rearrangements it undergoes in membrane fusion during the fertilization event, vesicle mediated transport or cellular locomotion [11, 13]. Such events integrate enormous structural change. It is the involvement of biological liquid crystals, or lipids, in these vital and dynamic processes that has caught the imagination of a sizeable body of researchers in the recent past and that motivates studies of lipid phase transition kinetics.

The stability of a given lipid phase depends on a range of biologically relevant thermodynamic quantities, both environmental and compositional [102, 131, 136)]. These include temperature, pressure, pH, salt and water concentration, and overall lipid composition. A change in any one or more of these can effect phase change, which in turn can profoundly influence the behavior, performance, and in some cases, survival of the system as a whole. Lipid phase transitions and the rates at which they occur are important for a host of other reasons. At the physiological level, they have been implicated in a range of processes including fat digestion [116, 136], development of disease [120, 136], regulation of membrane protein activity [3, 37, 38, 64, 69, 70, 78, 111, 113], membrane permeability and penetrability by macromolecules such as proteins [79], transbilayer lipid movement [35, 37, 39], visual transduction [3, 111], and anaesthetic action [77, 145]. It should be borne in mind that most membrane lipids in isolation have a variety of phases accessible to them at or close to physiological conditions [102, 131, 136], that many phase transitions can be effected isothermally [24, 27, 28, 102, 125, 131, 136, 144, 151], and that nonbilayer phases can accumulate in biological systems under conditions of stress [17, 25, 35, 36, 43, 56, 69, 70, 94, 125, 150].

From a commercial standpoint an understanding of lipid phase behavior is also important. For example, the organoleptic properties, shelflife and behavior during processing, formulation and reconstitution of foods, feeds and pharmaceuticals are sensitive functions of the phase relations existing among the various lipid components [87]. Cryopreservation and anhydrobiosis are other areas where membrane lipid phase behavior is important [17, 36, 53, 94, 137]. Failure to survive freezing and thawing protocols or extreme desiccation relates to mesomorph stability and to the dynamics and mechanism of phase changes taking place in the membrane lipid.

A study of lipid mesomorphism and the dynamics and mechanism of lipid phase transitions is interesting from a theoretical and mechanistic perspective [16, 31, 72, 133–135]. The thrust here is to decipher structural details of individual phases and of the local and macroscopic transformations between and within phases periodic in zero, one, two and three dimensions upon which are imposed the constraints of the hydrophobic effect. Lipids are liquid crystals. What is learned about these biological materials is of relevance to condensed matter physics.

In summary, there are many good reasons for studying lipid mesomorphism and the dynamics and mechanism of lipid phase transitions. The dynamic measurements enable limiting transition rates to be established and provide a basis for formulating, evaluating and refining transition mechanisms. Such measurements also reveal details of molecular structure and overall composition that influence phase stability and modulate transition rates and mechanism. The ultimate goal is to understand phase behavior and transition mechanism at a fundamental level and, thereby, effect control over lipid phase relations in vivo and in reconstituted, model and formulated systems.

1.2 Lipid Mesomorphism

Membrane lipids can exist in a number of intermediate states or mesophases more ordered than the liquid but less so than the crystalline solid [102, 131, 136]. Such materials are called liquid crystals and their multiple intermediate phase character is termed mesomorphism. There are at least four types of mesomorphic behavior: lyotropic, thermotropic, barotropic and ionotropic. These refer to the expression of the disparate liquid crystal phases by manipulating solvent (lyotrope) content, temperature, pressure, and salt concentration, respectively. Most membrane lipids exhibit the four types of mesomorphism.

Phase refers to that part of a system which is chemically and physically homogeneous (at a macroscopic level) throughout. The different lyotropic phases are distinguished by the relative spatial conjunction of lipid and solvent [95, 102, 131, 136]. In the case of the hydrated lamellar phase, for example, lipid molecules pack to form extensive planar bilayers which stack one atop the other, each separated by a layer of water. This phase has long and successfully been used as a model for the biomembrane. The lamellar phase is analogous to the smectic phase of the thermotropic liquid crystals and has many modifications arising, at least in part, as a result of differences in hydrocarbon chain packing [44, 136]. Packing can be crystalline (L_c), semicrystalline (L_c'), hexagonal or quasihexagonal with chains parallel (L_β) or tilted ($L_{\beta'}$) with respect to the bilayer normal. Chains may also be interdigitated (L_β^i), or if a periodic undulation exists in the lamellar plane, the phase is referred to as the ripple ($P_{\beta'}$) phase. In all of the above, chains are rigid and fully extended. Raising sample temperature can effect a chain order/disorder transition leading to a stabilization of the lamellar liquid crystal (L_α) phase.

Lipids can also adopt a number of nonbilayer configurations, the most commonly encountered of which are the hexagonal and cubic. The hexagonal phase consists of a hexagonally packed array of parallel rods of either water or lipid. In the normal hexagonal (H_I) phase, water constitutes the continuous medium supporting hydrocarbon rods, whereas in the H_{II}, or inverted hexagonal phase, hydrocarbon forms the continuous medium between aqueous rods. The cubic phase has many modifications that arise as a result of the three-dimensional packing of rodlike or lamellar aggregates, some of which have the property of forming infinite periodic minimal surfaces [71, 100, 103, 104]. It is emphasized that the nonbilayer phases are not simply physicochemical or crystallographic curiosities but are phases accessible to most membrane-derived lipids at or close to physiological conditions.

1.3 Time-Resolved X-Ray Diffraction

1.3.1 The Principle

A number of physical techniques are available for studying the dynamics of bulk lipid mesomorphic transitions. These include X-ray diffraction [1, 2, 15–23, 26, 28, 40, 42, 45, 88–91, 97–99, 117, 121–124, 129, 140, 142, 159, 160], Raman scattering [158], calorimetry [16, 29, 30, 32, 50, 79–81, 86, 105, 106, 108, 109, 117, 140, 155, 156, 157], dilatometry [5, 112, 118, 119], absorbance (turbidity, [10, 46, 48, 49, 55, 61, 62, 66, 74, 84, 115, 146, 147]), and fluorescence spectroscopy [54, 60, 73, 138, 143,

149, 156] polarized light microscopy [75, 126], pressure change [79–81], and electron spin [139, 148] and nuclear magnetic resonance [156]. A limitation of many of these methods, with the exception of X-ray diffraction, is that the measured parameter is oftentimes not a unique or a direct reflection of the phase under investigation. In contrast, X-ray diffraction provides direct phase identification and structural characterization in well-ordered systems. As an example, presented in Fig. 1 is a low-angle X-ray diffraction pattern of one of the cubic phases. The pattern is a structural fingerprint for the phase and, with sufficient resolution, uniquely identifies it. However, to

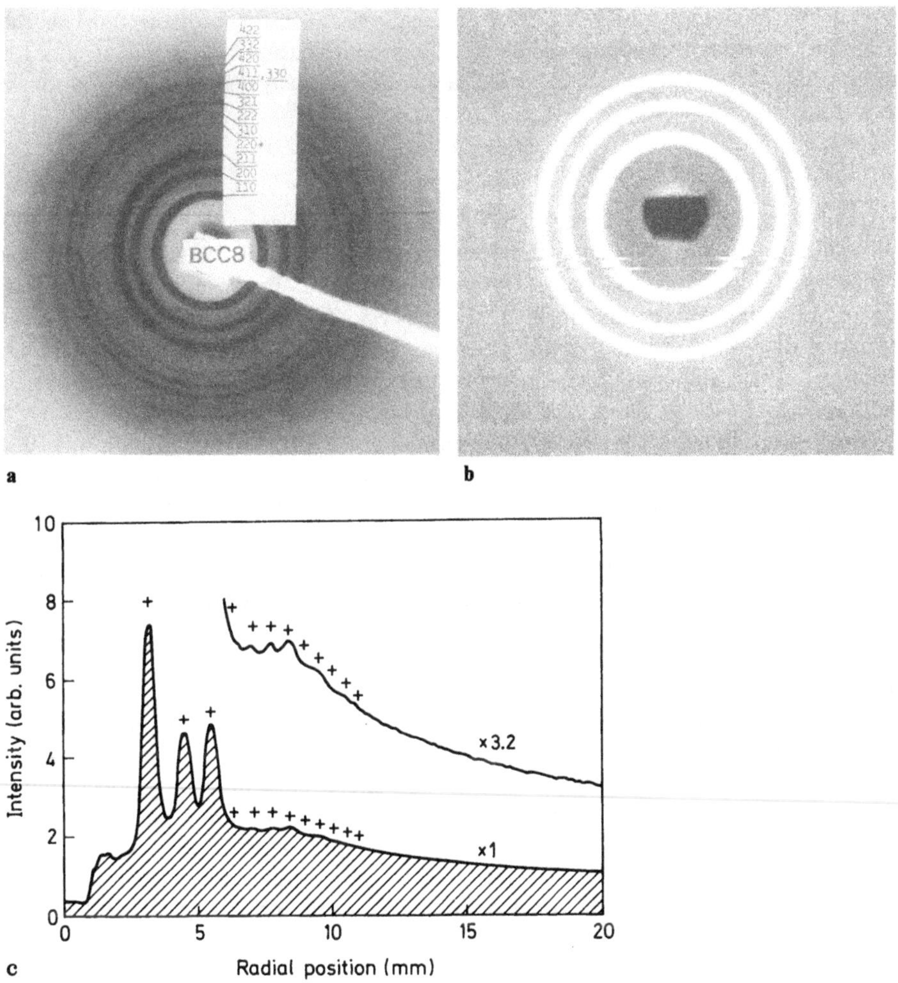

Fig. 1a–c. X-Ray diffraction patterns of the body centered cubic (Aspect No. 8) phase from hydrated monoolein recorded on X-ray sensitive film **(a)** and monoelaidin recorded on Kodak image plates **(b)**. A 360° radial average of the scattered X-ray intensity recorded in **(b)** is shown in **(c)** at relative magnifications of x1 and x3.2. Radial position is shown in units of mm corresponding to the actual distance on the image plate from the center of the pattern. The patterns in **(a)** and **(b)** were recorded with exposures of 15 min and 10 s, respectively

record such a pattern can take from minutes to days depending on the X-ray source, sample, phase type and detector. Accordingly, conventional X-ray diffraction techniques are not suited for examining the dynamics of lipid mesomorphism where phase changes can occur on a milliseconds time scale. This is where the method of time-resolved X-ray diffraction comes to the fore. The method represents a quantum leap, of sorts, because, with it, we can now directly and continuously monitor the dynamics and gain insights into the mechanism of structural rearrangements occurring as a lipid transforms from one mesomorph to another.

The ability to image X-rays continuously comes about as a result of the combination of two essential elements: on the one hand, a bright, high repetition rate source of (monochromatic) X-rays as is available at a synchrotron and, on the other, a live-time X-ray imaging device. This combination provides a means for direct phase identification and characterization continuously in time during the course of a transition.

1.3.2 Information Content

Insofar as the dynamics and mechanism of phase transitions are concerned, TRXRD provides two important pieces of information: 1) transition kinetics obtained by following the relaxation of the system in response to an applied perturbation and 2) details of intermediates that form in the process. These data provide a basis for formulating, evaluating and refining transition mechanisms (See Ref. [16] as an example). Since kinetics is the primary focus of this review, mechanistic aspects of lipid phase transitions will not be discussed here.

The TRXRD method is information-dense in the sense that an enormous amount of additional detail is provided in the course of the live-time measurements [16, 18, 89, 142]. Needless to say, the proposed mechanism must be consistent with such details. As an example, consider a TRXRD experiment conducted with a T-jump perturbation. The additional information gleaned from such an experiment includes: 1) the temperature dependence of the measured variable(s) in single and coexisting phase regions, 2) phase (and intermediate) lattice type and symmetry, lattice parameters. commensurability, 3) domain size, coherence length, disorder (in the form of diffuse scatter), 4) sample orientation, phase coplanarity, crystallinity, 5) reversibility, hysteresis in the transition temperature, phase type and lattice parameters, 6) continuous or discontinuous nature (order) of the transition, structural connectivity of adjacent phases, 7) temporality of changes in long- and short-range order, 8) cooperativity, thermal width of the transition, 9) lipid mixing, phase separations, and 10) temporal correlation of thermal events and structural rearrangements. Such a wealth of information supports the view that TRXRD is a powerful technique not only for establishing lipid phase transition kinetics but also for providing a useful insight into transition mechanisms.

1.3.3 Experimental Aspects

The experimental arrangement used by the author in making TRXRD measurements is shown schematically in Fig. 2. The essential components of this system are the synchrotron source, monochromating and focusing optics, the sample, the detector and recording and image analyzing devices [16, 18]. The former includes an image

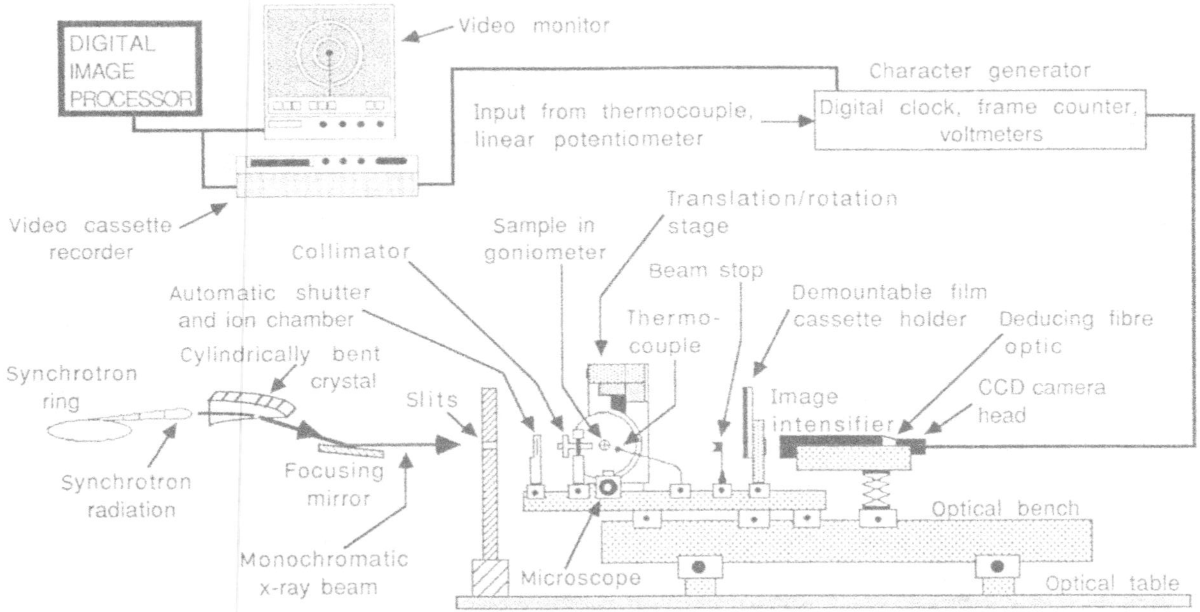

Fig. 2: Schematic diagram of the experimental arrangement for monitoring X-ray diffraction in live-time using synchrotron radiation (not drawn to scale). White radiation from the synchrotron is simultaneously monochromatized (λ = 1.57 Å, 7.9 keV) and horizontally focused by a 10 cm long, cylindrically bent germanium or silicon crystal. Higher order harmonic contaminants ($\lambda/3$, etc.) are eliminated and the 7.9 keV beam reflected and vertically focused by a 60 cm long nickel- or platinum-coated mirror. The monochromatic beam is focused to a point 1.5 mm wide $\times$ 0.2 mm high. X-rays diffracted from the sample are allowed to strike the fluorescent screen of a three-stage intensifier tube used for image intensification. The image is displayed dynamically on the intensifier tube, and recorded using a CCD camera, fiber optically coupled to the intensifier tube, and a video cassette recorder. Lipid samples in glass or quartz capillaries are mounted in a goniometer on the low-angle diffraction camera. A thermocouple to monitor sample temperature is placed either in or next to the capillary as close to the X-ray beam as possible. Along with the two-dimensional diffraction image, the experiment number, elapsed time in seconds and frame number, sample temperature and incident intensity are recorded simultaneously on video tape. Typical operating conditions at the Cornell High Energy Synchrotron Source are as follows: electron beam energy, 5.3 GeV; one, three or seven electron bunch mode with a total current of 20–80 mA; repetition rate in single bunch mode, 10^6 Hz; radiation pulse duration, 130 ps/electron bunch; total flux at the focal spot, 10^{12} photons s^{-1} (3×10^{10} photons delivered through a 0.3 mm diameter collimator)

intensifier tube, while the latter incorporates a charge coupled device (CCD), video camera, video cassette recorder and monitor and a digital image processor. Sample manipulation devices have been omitted from the figure for sake of clarity. It is emphasized that the system shown in Fig. 2 represents but a single configuration and that there are many more possible arrangements. Furthermore, the experimental arrangement is constantly being modified, reconfigured and upgraded. Flexibility in terms of equipment and experimental design is a critical aspect of successful TRXRD measurements.

1.3.4 Limitations

As a method, TRXRD is still in its infancy. While it has already proven to be a powerful, information-dense structural and kinetic probe there is much room for improvement. In what follows I draw upon my own experience as to the limitations of the method and will address these at four levels: facility, sample, instrumentation, and experimental design. The view held here is that deficits in the technique will be corrected more expeditiously if attention is drawn to them at an early stage. Alas, in some cases, a solution must await the development of new technologies. This exercise also serves to emphasize that TRXRD is not the panacea but rather another tool in an arsenal of physicochemical techniques with which to tackle critical issues in condensed matter phase science.

1.3.4.1 Facility

TRXRD measurements require items of high technology not the least of which is a synchrotron. Beam-time at any one of these national facilities is at a premium. Furthermore, measurements must be performed away from the comforts of the home institution usually on a 24 hours a day basis and under severe time constraints. Thus, experimental strategy and preparedness are integral to successful measurements of this type.

1.3.4.2 Sample

Lipid samples that exhibit well-ordered phases are optimally suited for TRXRD measurements. Lack of order suggests that other techniques such as electron or scanning tunneling microscopy may prove more useful for phase identification and characterization. Alternatively, the sample may be rendered more ordered by partial desiccation and/or preferential orientation [33].

Before proceeding with a kinetic analysis it is imperative to establish the equilibrium properties of the sample over the range of conditions to be imposed. This often amounts to constructing the corresponding T-(P)-composition phase diagram. Two methods (Sect. 3) based upon TRXRD have been developed recently which expedite acquisition of such data [23, 26].

Because of the enormous photon flux available at a synchrotron source, both thermal and nonthermal effects of X-radiation must be considered as potential problems when making TRXRD measurements. Nonthermal radiation damage has been shown to be significant in the case of hydrated phosphatidylcholine [15]. However, the degree of damage is lipid species specific at the least and each sample must be evaluated on an individual basis. A more complete description of the problem of X-radiation damage is presented in Sect. 2.

Chemical stability of the sample components must also be determined over the range of conditions to be studied during the course of a TRXRD measurement. Sample decomposition should be tested for upon completion of each experiment.

1.3.4.3 Instrumentation

a. Detectors

Much has been written about X-ray detection and the use of one- and two-dimensional detectors for TRXRD. Such issues as background, bloom, detector nonuniformity, dynamic range, linearity, sensitivity, size, spatial- and time-resolution, streaking, and so on are characteristics that have been reviewed quite extensively [6, 63, 65, 82, 83]. No attempt will be made to cover them here. In the ensuing paragraphs, an indication will be given of what might be expected as the quality of some of the aforementioned characteristics improves.

Time-Resolution. With improved time-resolution, shorter lived intermediates and faster transitions can be monitored. Since exposure time is inversely proportional to incident flux, faster measurements are increasingly realizable as synchrotron X-ray sources become brighter through enhanced emittance characteristics, use of insertion devices (wigglers and undulators) and improved X-ray optics. As an example of what is currently possible at existing sources, the reader is referred to Fig. 3. Included in the figure is a diffraction pattern recorded in live-time with a shutter-open time of 125 μs. Since the duty cycle of the machine is 10^{-4} (ca. 0.1 ns long pulses every μs) this amounts to a direct X-ray beam exposure of 12.5 ns. These data indicate that lipid X-ray diffraction measurements with millisecond-to-high microsecond time-resolution are now possible, provided suitable phase transition triggers can be implemented. Of course, with repeatable transitions, picosecond time-resolution is possible using a stroboscopic approach [92].

Sensitivity. Improving sensitivity means that low occupancy intermediates can be detected more readily and that data can be collected on more dilute, and possibly more physiologically relevant, membrane preparations such as unilamellar vesicles and whole cells. This advantage obtains only when scattering from the diluent does not interfere with that from the membrane component itself. Furthermore, with dilute suspensions rapid mixing (stop-flow and quench-flow) techniques can be put to good use. Such is not possible or can be achieved only with great difficulty with the highly concentrated and often viscous lipid preparations which must be used presently.

One- versus Two-Dimensional Detectors. For the most part, powder diffraction is the norm when dealing with lipid dispersions. Quantitative time-resolved powder diffraction measurements are possible with a one-dimensional detector only when randomly oriented samples are used and the powder character is preserved throughout the transition. Lipids have a propensity for preferentially orienting and for forming large crystallites in certain phases [18, 96]. In both instances, quantitation with a one-dimensional detector is likely to give rise to uninterpretable kinetics. In this regard

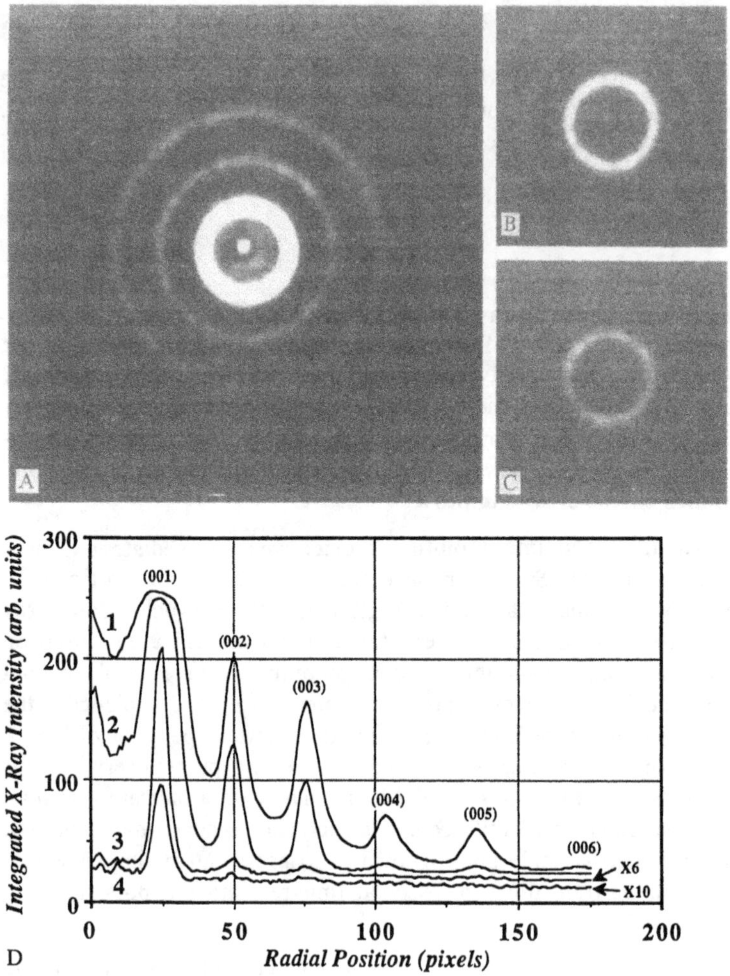

Fig 3A–D. Low-angle X-ray diffraction patterns of the lamellar crystalline phase of cadmium arachidate recorded in live-time with shutter open times of 20 ms **(A)**, 1 ms **(B)** and 125 μs **(C)**. Given a duty cycle of 10^{-4}, the latter correspond to actual X-ray beam exposures of 2 μs **(A)**, 100 ns **(B)** and 12.5 ns **(C)**. A 360° radial average of the X-ray intensity in each pattern is presented in **(D)**. Curves 1–4 in **(D)** correspond to shutter open times of 100, 20, 1 and 0.125 ms, respectively. The d_{001} spacing is 55.4 Å. Live-time images were recorded using the experimental arrangement shown in Fig. 2

the benefits of an area detector are obvious. In addition, radial integration of the two-dimensional powder pattern can be used to enhance the signal-to-noise ratio (see Fig. 1).

In the present context, it is worth noting that quantitation becomes more difficult if, in the course of a measurement, there is a change in the d-spacing of the relevant reflection. Such a change implies a sampling of different parts of the form or structure factor which alone, can effect large changes in scattered intensity (See Ref. [130] as a example).

Detector Size. To monitor simultaneously changes in long- and short-range order during a phase transition is a desirable feature in many TRXRD measurements. Unfortunately, most commercially available area detectors are of such limited size and spatial resolution that simultaneous low- and wide-angle measurements of the desired quality are not possible. Multiple detectors obviate the problem as does the less desirable option of making temporally distinct measurements in the low- and wide-angle region with a single detector. A large area (53 mm x 53 mm) CCD detector has recently become available commerically [76]. Its usefulness in continuous TRXRD measurements is limited by the need to operate it in a slow scan mode (ca. 1 min/frame) to reduce readout noise. However, as direct X-ray imaging devices, CCDs hold considerable promise especially since the smaller ones are reasonably priced [34].

Image Plates versus Film

A novel integrating detector referred to as an image or storage phosphor plate has recently become available and is presently being evaluated as an alternative to X-ray sensitive film [4, 8, 152]. In the course of recent experiments at the Cornell High Energy Synchrotron Source (CHESS) a comparison was made between the image plates and film for recording lipid dispersion powder patterns at 1.57 Å (Fig. 1). In terms of background, dynamic range and linearity the image plates excel. The difference in sensitivity is less obvious although this is expected to improve in favor of image plates at higher energies. As with film the image plate is an integrating detector of large size. However, upon processing of the plate direct digital data are obtained, unlike film which requires development and microdensitometry. The spatial resolution of the image plate (ca. 150 μm) is considerably lower than that of film [4]. Also, only a limited number of plates can be processed (15–20 min readout time) at any one time due to the large size and expense of the readout device. At this juncture, the evaluation of image plates as X-ray detectors is incomplete.

b. I_0 Characteristics

Beam instability and decay during the course of a time-resolved diffraction measurement are a constant source of concern at synchrotron facilities. For this reason a continuous measure of incident flux (I_0) during the course of an experiment is critical for subsequent data normalization.

c. Relaxation Methods

Kinetic analysis typically involves perturbing a system in one equilibrium state to another by the rapid adjustment (jump) of one or more thermodynamic variables or triggers. Relaxation of the system to the new equilibrium condition is followed by monitoring some observable which reflects rearrangements internal to the system which occur during the recovery period. The big advantage of TRXRD as a means for studying lipid phase transition kinetics is that the relative amounts of the two interconverting phases can be measured directly and quantitatively and that the individual phases can be structurally characterized throughout the transition.

In the present context, relaxation methods are of two types: environmental and compositional. The former includes temperature (T), pressure (P), and magnetic

Table 1. Characteristic times of lipid mesomorphic phase transitions established by time-resolved x-ray diffraction

Trigger[a]	Lipid[b, c]	Transition[d, e]	Direction[f]	Characteristic Time[g]	Ref.
Temperature					
-Fluid	DMPC	Lβ′-Pβ′	↑↓	>5–17 min	[2] h
		Pβ′-Lα	↑↓	<2–70 min	[1] h
	DPPC	Lβ′-Pβ′	↑↓	500 s–min/h	[88, 89]
		Pβ′-Lα	↑↓	<2–<1000 s	[88]
		L_{subgel}-Lβ′	↑	1.5–5 s	[142]
		gel-α	↑	2 s	[142]
		Lβ′-Pβ′	↑↓	≧1 h	[90]
		Lgel-α	↑↓	2–50 s	[22]
		Lc-α	↑↓	<20 s	[21]
	L-DPPE	Lc-Lα	↑	10 s	[142a]
	L, DL-DPPE	Lβ-Lα	↑↓	2 s	[142a]
	DHPE	Lβ′-Lα	↑↓	1–6 s	[16]
		Lα-HII	↑↓	1–5 s	[16]
		Lc-HII	↑	1–5 s	[16]
		L-HII	↓	4–5 s	[16]
		Lc-α	↑↓	1–6 s	[16]
		α-H	↑	2 s	[16]
	HOPE	Lβ-Lα	↑↓	2 s	[90]
		Lα-HII	↑↓	12 s	[90]
	PA	Lgel-Lα	↑↓	s–<1 min	[45]
	PE	L-H	NR[i]	<0.01 s	[91]
	MGDG	Lc2-α	↑	20 s	[99]
		Lc1-α	↑	3 s	[99]
		Lβ-α	↑↓	2–3 s	[99]
		Lc1-Lβ	↓	<6 s	[99]
		Lc2-Lc1	↓	d	[99]
		Lgel-HII	↑	25 s	[121]
		metastable gel-HII	↑	10 s	[121]
		Lc-Lβ	↓	8 min	[121]
		Lc1-Lβ	↑↓	30 s–20 min	[97]
	ME	Lgel-Lα	↑	1.9 s	[18]
		Lα-(X)-BCC8	↑	1.6 s	[18]
		BCC8-CP4	↑↓	0.5 s–min	[18]
		CP4-HII	↑↓	0.9–1.0 s	[18]
		HII-FI	↑↓	0.6–1.0 s	[18]
		CP4-FI	↑↓	0.9–1.2 s	[18]

	MO	CP4-HII	↑↓	0.6–1.5 s	[18]
		HII-FI	↑↓	0.1–1.5 s	[18]
		BCC12-CP4	↑↓	1.5–34 s	[18]
		BCC8-CP4	↑↓	0.8–300 s	[18]
	K Palmitate	(Lc1, Lc2, Lα1, Lα2, Lα3, Lβ]		NR	[98]
	Cholesterol Oleate	Lc-FI	↑↓	< 16–< 34 s	[21]
	Cholesterol linoleate	Crystalline-smectic	↓	4–8 min	[104a]
		Crystalline-FI	↑	< 4 s	[104a]
	POPC + HOPE	L-Micellar (Fusion)	↑	3 min	[90]
	DLPE + DPPC	Lgel-Lα	↑	1 s	[41]
	DMPE + DPPC	Lgel-Lα	↑	3 s	[41]
	DPPE + DPPC	Lgel-Lα	↑	9 s	[41]
	DSPE + DPPC	Lgel-Lα	↑	8 s	[41]
	E. coli Membranes	Lgel-Lα	↑	min–h	[45]
	DPPC + Trehalose	L_β-L_α	↑	2 s	[120a]
	L-DPPC + Trehalose	Lβ-Lα	↑↓	2 s	[o]
	DMPC	Lβ′-Lα	↑	1 s	[117a]
	+ butanol	$L_{\beta'}$-Lα	↑	4 s	[117a]
	+ octanol	$L_{\beta'}$-Lα	↑	4 s	[117a]
	+ dodecanol	$L_{\beta'}$-Lα	↑	4 s	[117a]
	+ tetradecanol	Lβ′-Lα	↑	1 s	[117a]
	DPPC + IM KSCN	L^i_{gel}-Lα	↑	1–2 s	[40]
	PA + Lysozyme	Lgel-Lα	↑	min–h	[45] h
-Peltier	Lipid extracts, Membranes of *E. coli*, *B. subtilis*	(L1, L2, L_{gel}, Lα, HII)	↑↓	2–5 s	[124]
	Egg PC	NR		ms–s	[123]
-Joule	DOPE	Lα-CP4-HII	↑↓	NR	[132]
	DOPE	Lα-HII	↑	0.01–100 s	[140]
-Laser	DEPE	Lα-HII	↑	0.01–100 s	[140]
	SOPE	L_β-L_α	↑	< 5 ms	[89a]
		L_α-HII	↑	> 0.2 s	[89a]
	DPPC	$L_{\beta'}$-$P_{\beta'}$	↑	< 10 ms–≫ 2 s	[l]
-Microwave		$P_{\beta'}$-L_α	↑	0.02 s	[l]
	DHPE	Lβ′-Lα	↑↓	2–6 s	[k]
		L_α-HII	↑↓	2–6 s	[k]
	DMPC	Lβ′-$P_{\beta'}$-Lα	↑↓	10 s	[k]
	ME	Lα-(x)-BCC8	↑↓	10 s	[k]
-NR		BCC8-CP4	↑↓	10 s	[k]
-Isothermal	1-O-alkylglycerol	α or subα-β		4–6 s	[117] h
	Fatty suppository bases	β′-like-α-like		8–624 h	[159, 160 h]
	DPPC	Lsubgel-Lβ′		min–d	[129] h

Table 1 (continued)

Trigger[a]	Lipid[b, c]	Transition[d, e]	Direction[f]	Characteristic Time[g]	Ref.
Pressure					
	DHPE	Lβ'-Lα	↑↓	3–32 s	[28]
	DMPC	Pβ'-Lα	↑↓	s–min	[j]
Composition					
Na Taurodeoxy-cholate	Egg PC	Lα-Micellar	↑	0.5 s–min	[88, 90]
Ca^{2+}	DOPS	Lc-Lα	↑	5 s	(This work)
H_2O	MO	(FI, Lα, BCC12, CP4)	↑	variable	[20a, 55a] m
	DMPC	(Lc, Lβ, Lα)	↑	min	[n]
Li^+	DOPS	Lc-Lα	↑	min	[n]
H^+	DMPC	gel-α	↑	min	[n]

[a] For thermally induced phase transitions adjustment in sample temperature made use of a flowing fluid (air or liquid), a Peltier or thermoelectric device, Joule (capacitance) or microwave heating as indicated.

[b] Individual references must be consulted for experimental details including sample composition.

[c] Lipid abbreviations are as follows: DHPE, dihexadecyl PE; DLPE, dilauryl PE; DMPC, dimyristoyl PC; DOPE, dioleoyl PE; DOPS, dioleoylphosphatidylserine; DPPC, dipalmitoyl PC; DSPC, distearoyl PC; HOPE, 1-o-hexadecyl-2-oleoylphosphatidylethanolamine; ME, monoelaidin; MGDG, monogalactosyldiacylglyceride; MO, monoolein; PA, phosphatidic acid; PC, phosphatidylcholine; PE, phosphatidylethanolamine; SOPE, 1-stearoyl-2-oleoyl-sn-3-phosphatidylethanolamine.

[d] The transitions are listed with the low temperature phase to the left of the hyphen.

[e] The lipid phase notation is as follows: BCC12, body-centered cubic, aspect No. 12 (Ia3d); BCC8, body-centered cubic, aspect No. 8 (Im3n, 1432, 143m, Im3, I23, $I2_13$); CP4, cubic primitive, aspect No. 4 (Pn3m, Pn3); FI, fluid isotropic or melt; HII, inverted hexagonal; Lα, lamellar liquid crystal; $L_\beta(L_\beta')$, lamellar gel with rigid chains parallel (or tilted) to the bilayer normal; Lc, lamellar crystalline with variants denoted by numerical subscripts; Lgel, lemellar phase with rigid chains not packet in a crystalline arrangement; P_β', ripple phase; X, an, as yet, unidentified intermediate.

[f] Direction refers to whether the intensity or quantity of the applied variable (temperature, pressure, concentration) was increased (↑) or decreased (↓).

[g] Characteristic time is used in the most general sense and includes time to complete the transition, half-time of the transition, and so on. The values cited cover the range reported as a function of many variables including the direction of the transition. Obviously, individual References must be consulted for a complete explanation.

[h] These measurements were made using conventional x-ray sources.

[i] NR implies the quantity was either not reported or could not be deciphered from the available data.

[j] Caffrey, M., Mencke A., unpublished observations.

[k] Caffrey, M., Magin, R., Hummel, B., Zhang, J., unpublished observations.

[l] Laggner, P., unpublished observations.

[m] see Fig. 7.

[n] Caffrey, M., unpublished observations.

(B), electric (E) and gravitational field strength (G), while the latter encompasses changes in lipid species type and mole fraction, as well as adjustments in proton, salt, physiologically relevant cation, water, and other additive concentration.

To date, the most extensively used trigger has been temperature, probably because of the ease with which it can be applied (See Table 1 and references therein). T-jump can be effected by resistance (Joule), laser and microwave heating, and through the use of temperature-regulated fluid (liquid and gas) streams. Each method has its own advantages and disadvantages which, for lack of space, cannot be gone into here. The requirement in each case is to transfer heat rapidly and uniformly throughout the sample. This need applies both in adjusting sample temperature from the initial to the final value and in supplying and removing the latent heat of the transition. Measurements and calculations show that heat transfer can be limiting and can result in sizeable gradients and nonuniform thermal lags in the sample (see Ref. [16] and Fig. 4). One such calculation shows that it takes 1.3s to supply the latent heat ($\Delta H = 7.9 \text{ kcal mol}^{-1}$) of a transition occurring at 66 °C in a 1 mm diameter X-ray capillary when a temperature regulated air stream is used to effect a T-jump from 30 °C to 92 °C. Similarly, 0.4s are required to raise the temperature of the entire sample volume through the transition range of 1 °C. These calculations indicate the limitation of conductive heat transfer. They also highlight approaches for effecting more rapid T-jumps which include the use of 1) a fluid with a higher heat transfer coefficient than air, 2) smaller diameter capillaries which amounts to using thinner samples, and 3) higher fluid flow rates.

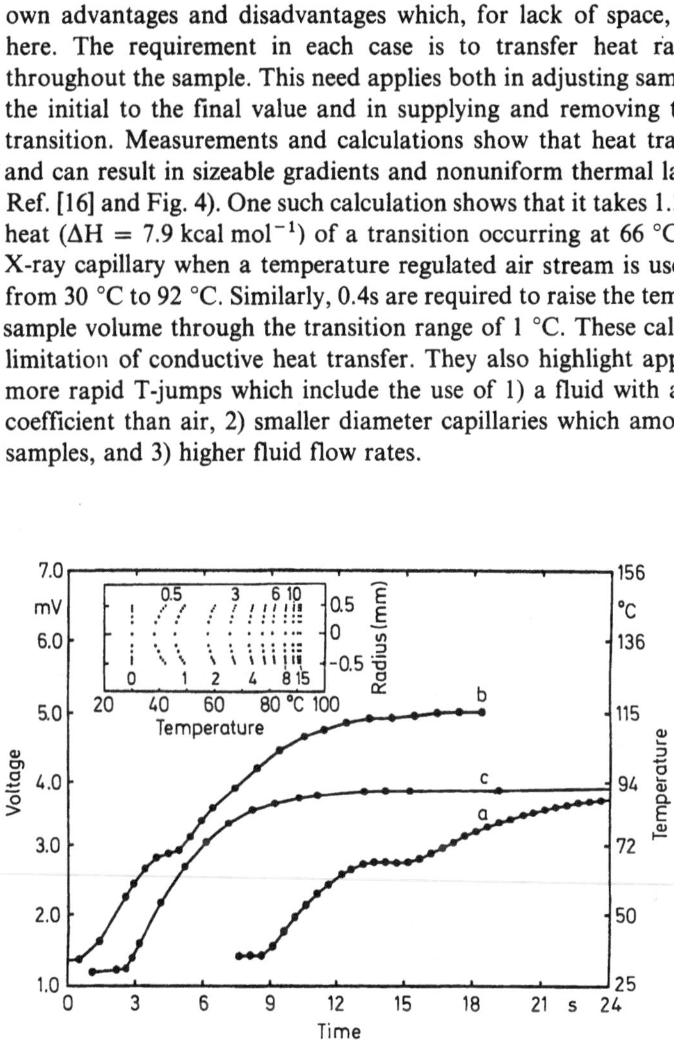

Fig. 4. In-sample temperature change with elapsed time after the commencement of heating hydrated phosphatidylethanolamine **(a, b)** and water **(c)** samples contained in 1 mm diameter capillaries using a temperature-regulated coaxial air stream. The thermal lag in **(a)** and **(b)** is due to the diversion of heat away from raising sample temperature and into chain melting which accounts for most of the enthalpy of the transition at 66 °C. T-jumps were from 30 °C to 92 °C **(a, c)** and 125 °C **(b)**. The inset shows the calculated temperature profile across the capillary diameter in **(a)** as a function of time in seconds following the T-jump. Adapted from Ref. [31]

This same reasoning applies to transitions brought about by any of the other triggers noted above. For example, recently performed P-jump measurements on the $L_{\beta'}/L\alpha$ transition in a hydrated phospholipid depicted in Fig. 5 suggest that the conduction of the latent heat of the transition away from the sample may control the conversion rate [28].

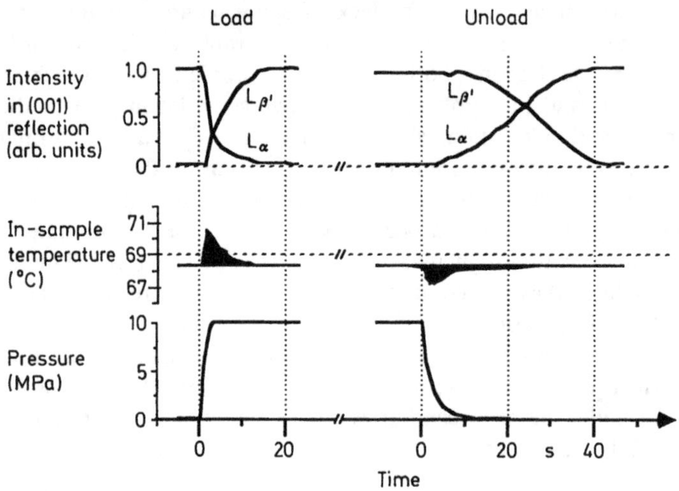

Fig. 5. Progress of the pressure-induced $L_{\beta'}/L$ phase transition in hydrated phosphatidylethanolamine monitored by time-resolved X-ray diffraction. Included in the figure is the changing scattered X-ray intensity in the (001) lamellar reflection, pressure and in-sample temperature following a 9.64 MPa (96.4 atm, 1400 psi) pressure-jump applied in the load and unloading directions. The data clearly illustrate a recurring limitation in many of these measurements, namely, the control of the transition by heat flux into and out of the sample. This is shown in the load curve. However, heat flow need not always be a limitation as is evident in the unload curve. (Unpublished observations, M. Caffrey and A. Mencke)

Because of the need to work with concentrated lipid dispersions conventional rapid mixing techniques have not been commonly used in TRXRD measurements to date. Further, when working with complex lipid aggregates such as multilamellar vesicles, the rapid and uniform distribution of an additive throughout the sample is difficult to achieve because of the need to pass through successive, concentric layers of lipid and water. Photolyzable caged initiator complexes [107] might be used to advantage here when appropriate and when available. Despite the problems, a limited number of compositional jump experiments have been performed. For example, the Ca^{2+}-induced disorder/order phase transformation was examined in hydrated phosphatidylserine dispersions following a jump from 0 to 0.5 M $CaCl_2$ (Fig. 6, Table 1). As judged by temporal changes in low-angle X-ray diffraction from the lipid, the transition is 50% complete in 5 s. In a similar way, it has been possible to examine pH, salt and water induced phase changes in a variety of lipid systems (Table 1).

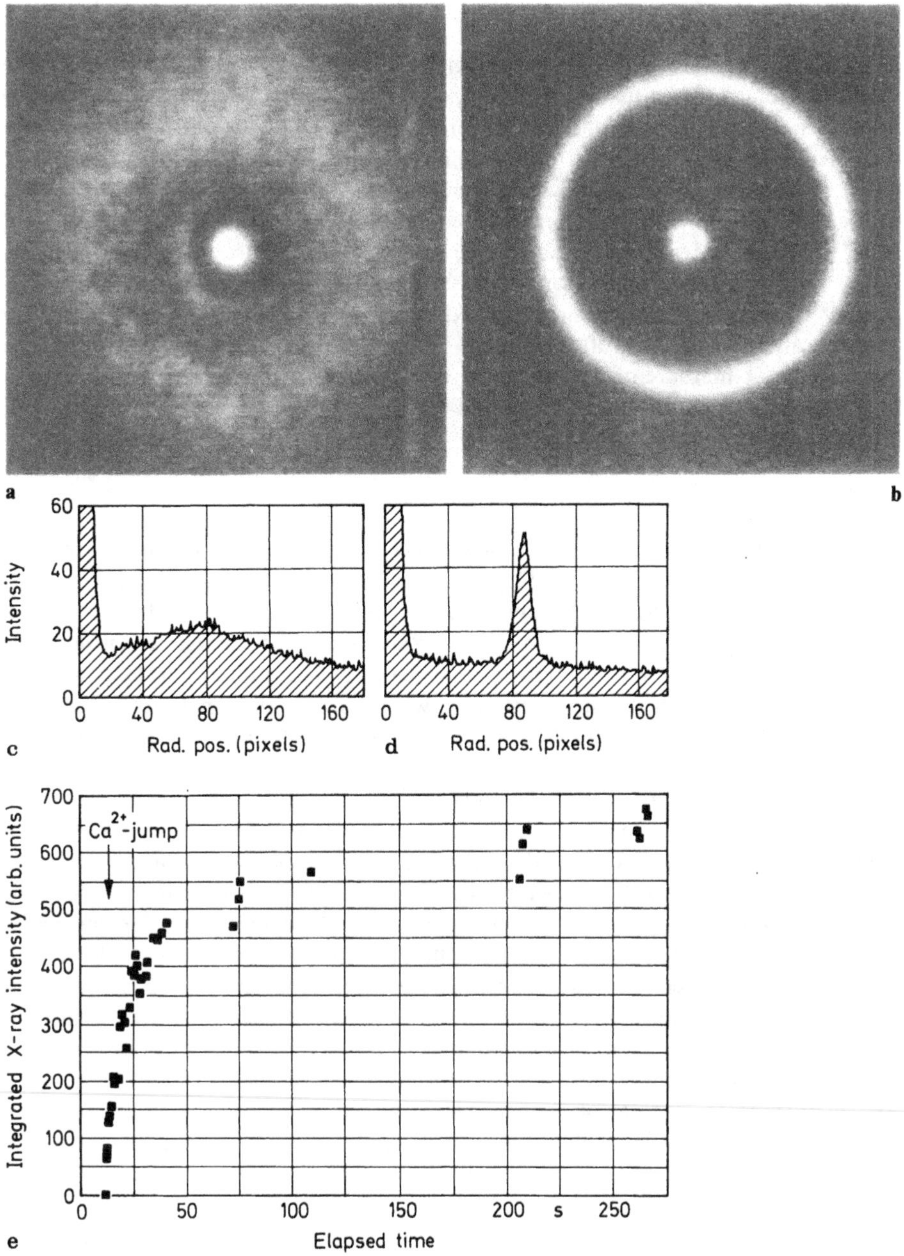

Fig. 6a–e. Kinetics of a Ca^{2+}-induced phase transition in fully hydrated phosphatidylserine monitored by low-angle time-resolved X-ray diffraction. Photographs of the live-time images on the video screen before and after the 0 to 0.5 M Ca^{2+}-jump are shown in **(a)** and **(b)**, respectively. Radial averages of the scattered X-ray intensity in **(a)** and **(b)** are presented in **(c)** and **(d)**, respectively. Radial position is shown in units of pixels (picture elements) from the center of the pattern as recorded on the video image. In **(e)** is shown the progress curve of the Ca^{2+}-induced transition where intensity corresponds to peak area of the sharp (001) reflection at ca. 51 Å of the highly condensed lamellar crystalline phase

1.3.4.4 Design

For TRXRD measurements to be meaningful, special attention must be devoted to the questions posed, to the manner in which the experiments designed to answer these questions are performed, and to the proper choice of materials and complete documentation of experimental protocols and results. As an example, in the case of a T-jump experiment, documentation should include direction, rate and magnitude of the jump, initial and final temperatures, and the thermal history of the sample. Ideally, in-sample temperature should be recorded at or close to the interrogating X-ray beam throughout the course of the transition. The manner in which the characteristic transition time is determined must be adequately described and, where possible, an estimate made of the number of molecules transforming per unit of time and volume. A measure of the sensitivity of the detection system to the presence of intermediates or minor and disordered phases would enormously benefit data evaluation.

1.3.5 Results to Date

The first TRXRD study of phase transitions occurring in membranes and membrane lipid extracts was described in 1972 [45]. It is interesting to note that these time-resolved measurements were realized through a technological innovation in the form of a linear, position-sensitive proportional X-ray counter. In the past decade, considerable interest in lipid phase transition kinetics has developed in response to the emergence of new technologies, the most important of which include the synchrotron radiation source, and X-ray optics and detectors. The increased interest level is reflected in a growing literature.

A summary of the results to date pertaining to lipid phase transition kinetics established by TRXRD is presented in Table 1. Included in the table are data obtained using synchrotron X-radiation and a sampling of data collected using conventional X-ray sources. Entries are arranged by the type of trigger used to effect the transition, which includes temperature, pressure and composition. The means by which T-jumps were implemented include fluid flow (air or liquid) around the sample or sample compartment, as well as Peltier, Joule (capacitance) and microwave heating.

Approximately twenty different lipid systems have been examined by the method of TRXRD. The list includes single lipid species as well as binary lipid, lipid-small molecule and lipid-protein mixtures and isolated membranes and membrane lipid extracts (Table 1). A total of thirty, nominally distinct phase transition types are described in the table. The actual number of disparate phase transitions is expected to be less than this because of redundancies arising from an inadequate nomenclature and/or incomplete phase characterization.

The parameters describing the kinetics of lipid phase transitions have been reported in the literature in a variety of ways. This includes transit time or time to completely convert from one phase to the next, relaxation time, half-time, and so on. The relevant values are assembled under the general heading of "characteristic time" (τ_c) in Table 1 to convey a sense of the relative orders of magnitude. The range of τ_c values in the table spans nine orders from milliseconds to months. The lower limit of 10–100 ms perhaps reflects the limiting resolution of the X-ray detection and recording devices and the inability to trigger the transition rapidly and uniformly

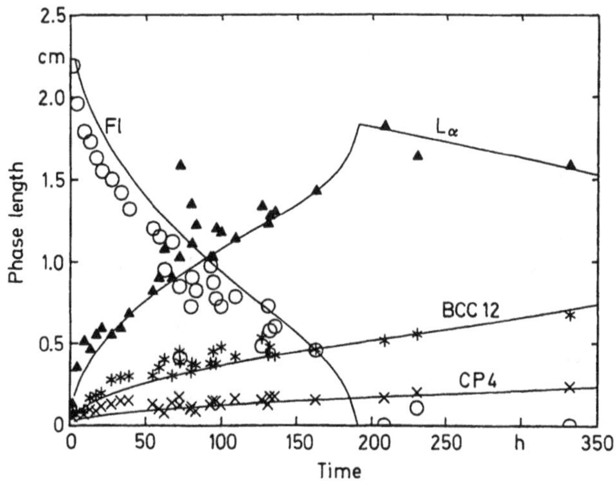

Fig. 7. The dynamics of monoacylglyceride hydration in the absence of mechanical mixing monitored by time-resolved X-ray diffraction. Mesomorph lengths along the lyotrope gradient are shown as a function of elapsed time following initial contact of an excess of pure water with a 2.5 cm long plug of pure monoolein (C 18:1c9). The continuous lines represent a fit to the experimental data based on a finite difference diffusion simulation [55a]. The water diffusion coefficients obtained from the simulation are 1.0, 1.4, 1.5 and 0.7×10^{-6} cm^2 s^{-1} in the fluid isotropic (FI), lamellar liquid crystal (L_α), cubic, body-centered (space group 12, BCC12) and cubic, primitive (space group 4, CP4) phases, respectively

more so than the intrinsic transition rate. The bulk of the τ_c values in Table 1 fall within the time domain of seconds. Once again, this clustering may reflect instrumental limitations more so than the intrinsic transit times. It is important to emphasize that the τ_c values reported include the time required to supply or remove the latent heat of the transition and to bring the entire sample through the full transition range as well as the intrinsic transit time [16, 18]. As a result, τ_c values represent upper bounds. These effects are indicated clearly in Fig. 5.

The data in Table 1 were assembled more in the spirit of presenting an overview of what has thus far been accomplished and less in pursuit of concrete conclusions. Such an expectation might be considered premature at this stage because of the limited size of the data base, and because the particulars of each entry in the table cannot be addressed fully in a review of this scope. Despite these limitations, some general observations will be made in regards to the measurements presented in Table 1.

- Some lipid phase transitions are fast, occurring on a time-scale of milliseconds, while others are extremely sluggish requiring weeks to complete.
- Certain transitions are equally fast in the forward and reverse direction. Others have rates that differ by many orders of magnitude.
- Many of the transitions studied appear to be two-state in the sense that at any point during the transition only two phases co-exist [16, 18, 89]. In contrast, other transitions are continuous, displaying a pronounced second-order character [142] while others again reveal the presence of transition intermediates [18].

- While some transitions are readily reversible and nonhysteretic in phase type, lattice parameters and transition temperature, others are extremely hysteretic [16, 18, 89, 142].
- For certain lipid systems, τ_c is independent of hydration, salt concentration, lipid identity, changes in the periodicity of the transforming phases and in long- and short-range order but is dependent on the magnitude and direction of the T-jump [16, 18] and the identity of the small molecule additive [117a].
- Long-range order is preserved throughout many lipid phase transitions [16, 18, 89]. This suggests that the transforming units remain coupled and undergo the transition cooperatively and that long-range order is established rapidly in the nascent phase. In contrast, other transitions involve an intermediate state devoid of long-range order. In such cases, the precipitous loss of the (ordered) phase undergoing change is accompanied or followed by the emergence of diffuse scatter and/or line broadening which sharpens up with time and eventually gives way to the ordered nascent phase [16, 18, 89, 140].
- Depending on the system, temporal correlations of changes in long- and short-range order may [142] or may not [16, 129] exist.
- In many of the transitions studied thus far, the interconverting phases appear to be incommensurate [16, 18]. This may reflect a bias toward the use of such systems which, by virtue of having distinct and well-resolved diffraction patterns, are more easily characterized.
- In certain systems, changes occur in short-range order without a corresponding adjustment in long-range order of the interconverting phases [142].
- In addition to rearrangements occurring during a phase transition, quite extensive and rapid changes take place within single phase regions. For example, the cubic phase unit cell volume in hydrated monoolein shrinks at a rate of 33% of its initial value per second during the course of a (66 °C) T-jump [18]. This implies facile lyotrope transport throughout the three-dimensional cubic network.

Despite the newness of the TRXRD technique, the data presented in Table 1 and related observations show that a sizeable body of information has been gathered concerning the dynamics and mechanism of lipid phase transitions. Further, the data serve to illustrate the potential of this approach in revealing the molecular structure and compositional dependence of mesomorph stability and interconvertibility and, ultimately, the underlying transition mechanism.

1.4 A View to the Future

TRXRD represents an important recent innovation in the experimental study of bulk lipid phase transition kinetics. The method provides direct structural information continuously throughout the course of the transition and offers useful insights into the transition mechanism. Although several successful experiments have been performed, the full potential of the method has yet to be realized. As indicated above, many of the limitations are of a technical nature. Next, I address some of the factors which, in my opinion, warrant attention if these limitations are to be removed and the potential of the TRXRD method fully exploited.

As far as incident photon flux is concerned the future looks "bright". Given the progress that has been made in the area of synchrotron beam emittance, insertion devices and X-ray optics, it is unlikely, in the near future at least, that TRXRD measurements will be limited by the available photon flux. More critical at this juncture are those elements in the TRXRD system "downstream" from the X-ray source.

The first of these elements is sample related. At issue here is the manipulation of sample environment and composition. A special effort must be devoted to developing new approaches for effecting rapid and uniform changes in sample environment and composition, singly and in combination. Additionally, methods must be devised for monitoring directly the applied thermodynamic variable(s) in the sample at or next to the interrogating X-ray beam at the same time as the TRXRD data is being collected. Another direction worth pursuing is the use of complementary methods in combination. Envisioned here is the simultaneous measurement of TRXRD and any one or more of the following: differential scanning calorimetry, polarized light microscopy, electron spin resonance, refractive index, and infrared and Raman scattering. In this way, structural changes determined by TRXRD can be correlated directly with whatever other parameter(s) is(are) monitored. The information gained using the combination approach far outweighs that obtained by performing each separately in time. In those cases where thermal history influences subsequent behavior the use of methods in combination should prove particularly beneficial.

The use of more dilute lipid dispersions, unilamellar vesicles and, ultimately, cell suspensions is another direction worth pursuing. Such possibilities become more realistic as incident photon flux increases and more sensitive X-ray detectors become available. Since conventional rapid-mixing techniques can be used with more dilute suspensions, an enormous benefit will accrue in studies where the kinetic event is triggered by a jump in sample composition.

While the contents of Table 1 might suggest than many of the transition of interest have already been examined, realistically it represents but the tip of the iceberg. What is needed now is a program designed to characterize systematically all known lipid phase transitions at the level of: 1) establishing limiting transit times, 2) characterizing phase transition intermediates, 3) deciphering transition mechanisms and 4) identifying critical aspects of molecular structure and sample composition that influence transition kinetics and mechanism. Particular emphasis must be placed on establishing limiting transit times for the faster transitions and on characterizing short-lived, low-occupancy and less ordered transition intermediates.

Concerning the detector, the author admits a bias toward two-dimensional imaging for reasons noted earlier. However, a combination of a one- and a two-dimensional detector system may prove advantageous also. The former would provide the rapid time-slicing while the latter could be used to monitor the powder character of the transforming sample and for subsequent radial averaging of scattered X-ray intensity. There is a pressing need also for larger area detectors such that time-resolved diffraction data can be recorded with sufficient spatial resolution in the low- and wide-angle regions simultaneously. The possibility of using CCDs as direct X-ray imagers and the image or storage phosphor plate as an alternative to X-ray sensitive film must be examined more fully.

Another important issue in this area of TRXRD is the ability, and indeed the discipline, to review immediately recently acquired data. When working at a synchrotron, the time constraints are such that there is a tendency to spend most, if not all, of the available time in data collection — the fear being that the synchrotron may fail at any moment. This approach can often give rise to inferior quality data or incomplete data sets. The opportunity to repeat the experiment and to make good such deficits may be many months away by which time other experiments have assumed precedence. The data collection and analysis system must have a built-in "instant" replay feature which should be extensively used.

Finally, in the interests of making most efficient use of this valuable resource, synchrotron radiation, serious consideration should be given to establishing close contacts between the experimentalist and the theorist. Such collaboration is viewed in the same spirit as the relationship existing between the experimentalist and the statistician. In the latter case, the experiment is designed ahead of time with input from both parties in such a way that when the data is collected a proper statistical analysis can be performed. Similarly, TRXRD data must be collected with a view to testing a particular hypothesis or addressing a well formulated question. Given the difficulties involved in making these measurements it makes sense to devote special care at the design stage so as to ensure that the maximum benefit is derived from each set of data and that wasteful duplication is avoided.

1.5 Conclusions

The potential of TRXRD as a means for establishing the kinetics and deciphering the mechanism of lipid phase transitions has been demonstrated. To date, only a fraction of the total number of lipid phase transitions has been examined using this approach. In many instances, the measurements have been cursory. Accordingly, there exists an amount of work yet to be done with the goal of establishing limiting kinetics, deciphering mechanism and determining how molecular structure and sample composition influence kinetics and mechanism. In the course of these measurements special attention must devoted to testing the potential of the TRXRD method with simple, well-behaved lipid systems. A judicious choice of sample and experimental protocol along with a prior knowledge of the equilibrium properties of the system will contribute in great measure to the success of the time-resolved studies. Despite the recent origin of the TRXRD technique, it is imperative now to push the method to its limits in many directions while at the same time enhancing its capabilities by implementing novel design and flexibility features and by incorporating new technologies.

2 Lipid X-Radiation Damage

While synchrotron radiation offers many advantages for X-ray diffraction studies of lipids it also has its severe limitations. In this regard we find adherence to one of the basic tenets of quantum mechanics, namely, that simply observing a system perturbs the system. The perturbation in question is radiation damage. Heretofore, radiation damage has not been a problem in the study of model and biological membranes by X-ray diffraction when conventional X-ray sources are used. With the advent of synchrotron radiation providing orders of magnitude more intensity on the sample, radiation damage has been shown to be a real problem [15]. Further, the inclusion of insertion devices (wigglers, undulators) along with improved X-ray optics will provide an even greater flux on the sample and the problem of radiation damage is expected to worsen.

In the aforementioned study, radiation damage of hydrated lecithin membranes brought about by exposure to wiggler-derived synchrotron radiation at 8.3 keV (1.5 Å) is reported. Considerable damage was observed with exposures under 1 h at an incident flux density of 3×10^{10} photon s^{-1} mm^{-2}, corresponding to a cumulative radiation dose of $\leqq 10$ MRad. Damage was so dramatic as to be initially observed while making real-time X-ray diffraction measurements on the sample. The damaging effects of 8.3 keV X-rays on dispersions of dipalmitoylphosphatidylcholine (DPPC) and phosphatidylcholine (PC) derived from hen egg yolk are as follows: (1) Marked changes were noted in the X-ray diffraction behavior, indicating disruption of membrane stacking. (2) Chemical breakdown of lecithin was observed. (3) The X-ray beam visibly damaged the sample and changed the appearance of the lipid dispersion when viewed under the light microscope.

Calculations show that a flux of 10^{12} photons s^{-1} mm^{-1} incident on a 1 mm thick sample effects an adiabatic temperature rise of approx 0.2 °C s^{-1} [15]. In the present case, it appears as though radiation damage is more likely a result of the ionizing rather than the heating effects of 8 keV X-rays. However, with the expected increases in incident flux such a thermal effect may become limiting.

Considering the importance of X-ray diffraction as a structural probe and the anticipated use of synchrotron radiation in studies involving membranes, the problem of radiation damage must be duly recognized. Furthermore, since DPPC, the major lipid used in the present study, is a relatively stable compound, it is not unreasonable to expect that X-ray damage may be a problem with other less stable biological and nonbiological materials. These results serve to emphasize that whenever a high intensity X-ray source is used, radiation damage can be a problem and that the sensitivity of the sample must always be evaluated under the conditions of measurement. Simultaneous translation and rotation of the sample during an X-ray exposure along with the judicious use of a beam shutter will minimize the effects of radiation damage.

3 Two New Approaches for Studying the Mesomorphic Properties of Lipids

As noted above, lipids exist in a number of intermediate physical states or mesomorphs between the crystalline solid and the isotropic liquid. The stability of these phases depend on temperature and composition and each lipids' pattern of dependency is conveniently described in the form of an isobaric temperature-composition phase diagram. Over the past few years two new and related methods of collecting mesomorphic phase information which are less time-consuming and more efficient and informative than conventional techniques have been developed. By incorporating a range of conditions into each sample preparation — a temperature gradient in the first method and a lyotrope gradient in the second — and utilizing the method of TRXRD, the time required to collect phase information for a complete diagram is reduced to minutes. Temperature-composition phase diagrams constructed using these methods compare well with those constructed by conventional means.

Lipid phase diagrams conveniently summarize mesomorph-temperature-composition dependencies (vide infra, Fig. 9, 11). With temperature on the ordinate and composition on the abscissa, the diagrams indicate the phases that exist over the entire range of conditions. Vertical lines in the diagram, which show the phases that exist for a sample of particular composition over a range of temperatures, are referred to as isopleths. Horizontal lines, which show the phases that exist for a range of compositions at a particular temperature, are referred to as isotherms.

In the traditional approach to collecting lipid mesomorphic phase data, a series of samples encompassing the entire range of composition to be analyzed are raised in a step-wise manner to progressively higher temperatures at each of which the phase is identified. This is referred to as the equilibrium, isoplethal method. It is isoplethal because it involves moving up isopleths in the phase diagram and is an equilibrium method because samples of fixed composition are equilibrated at discrete temperatures so that they represent particular combinations of temperature and composition — single points on a phase diagram.

The equilibrium, isoplethal method has a number of drawbacks [92a]. First of all, it is time-consuming. A number of samples must be raised to progressively higher temperatures and equilibrated for a time at each temperature before being separately analyzed for phase type. Secondly, vertical boundaries are difficult to isolate. And thirdly, because only a limited number of samples representing isolated points in the diagram are analyzed it is possible that phases which exist over a small range of conditions will go undetected.

3.1 The Underlying Principles

To overcome the limitations of conventional equilibrium techniques, two new approaches utilizing TRXRD have been developed to expedite lipid phase data acquisition [20a, 23, 26]. The underlying principle in both methods is that a continuous range of conditions, a gradient of either temperature or composition, is incorporated into each sample. The samples are placed in X-ray capillary tubes, and the gradients are established along their lengths. Thus, each sample represents not a single point in a phase diagram but an entire line.

In the first approach, a temperature gradient is imposed on samples of constant composition so that each sample represents a vertical line, or isopleth, in the corresponding phase diagram. In the second approach, the sample contains a solvent gradient and is raised to progressively higher temperatures. At each temperature the sample represents a horizontal line, or isotherm, in the phase diagram. These techniques are referred to, respectively, as the temperature gradient and lyotrope gradient methods.

The sample preparations create the potential for more efficient phase data collection, but making the most of this potential requires a means of rapid phase identification and characterization. In this regard, X-ray diffraction is a particularly powerful technique because it provides *direct* structural information. Since each molecular arrangement characteristic of a given mesomorphic phase generates a unique diffraction pattern, diffraction data provide the means to positively identify phases in single phase regions, to determine the proportion of each phase type in multiphase regions, and to structurally characterize each phase. However, the low photon flux of conventional X-ray sources necessitates long (hours to days) exposure times to produce satisfactory diffraction patterns.

This time limitation is overcome by using TRXRD. To analyze samples in both the temperature and lyotrope gradient methods, capillaries are passed through a

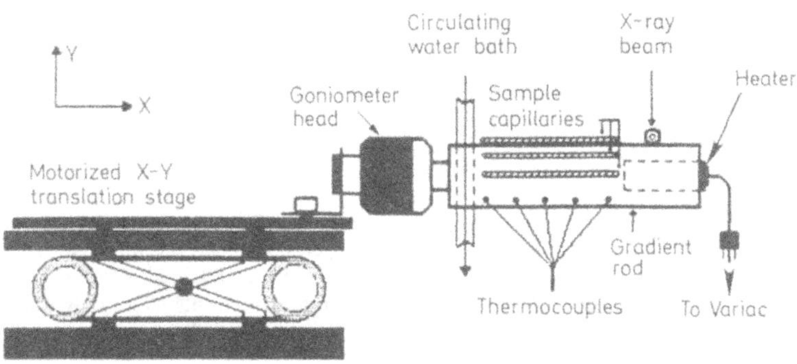

Fig. 8. Schematic diagram of the experimental arrangement used in making live-time X-ray diffraction measurements with the temperature gradient apparatus

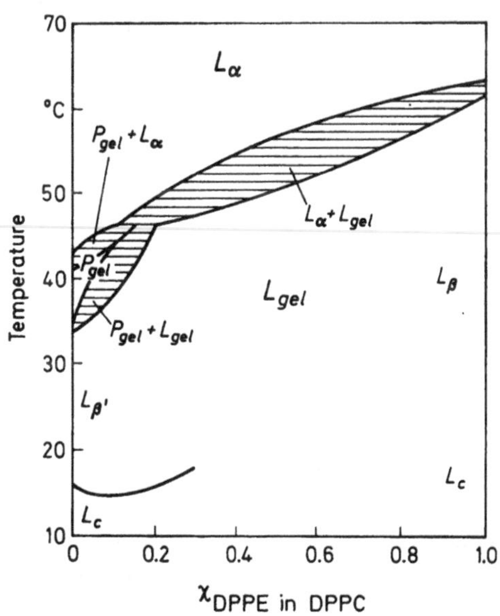

Fig. 9. Temperature-composition isobaric phase diagram for the fully hydrated dipalmitoylphosphatidylcholine/dipalmitoylphosphatidylethanolamine system constructed using the temperature gradient method. The notation used is that of Luzzati [8] and is as follows: L_c, lamellar crystalline (also referred to as the subgel phase); L'_β, lamellar gel phase with hydrocarbon chains tilted with respect to the bilayer normal; P_{gel}, ripple phase; L_α lamellar liquid crystal phase. Insert bold line in graph as indicated

synchrotron-derived X-ray beam while a video camera records the image-intensified diffraction pattern in live-time continuously along the length of a sample. A translation stage positioned perpendicular to the X-ray beam moves the sample capillary tube through the beam. With this system it is possible to record diffraction information in live-time continuously along the length of the sample capillary (Fig. 8 and 10).

The diffraction patterns indicate the phases present in the sample and the relative amount of each phase in multiphase regions. By recording capillary position along with the changing diffraction pattern on video tape, phase boundaries, signified by changes in the diffraction pattern, are assigned to precise points in the sample. In the temperatures gradient method, each point corresponds to a specific temperature,

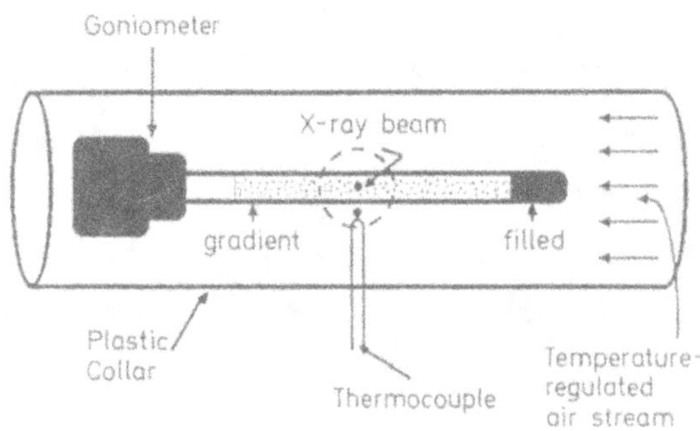

Fig. 10. Schematic diagram of the experimental arrangement used in making time-resolved X-ray diffraction measurements with the lyotrope gradient apparatus

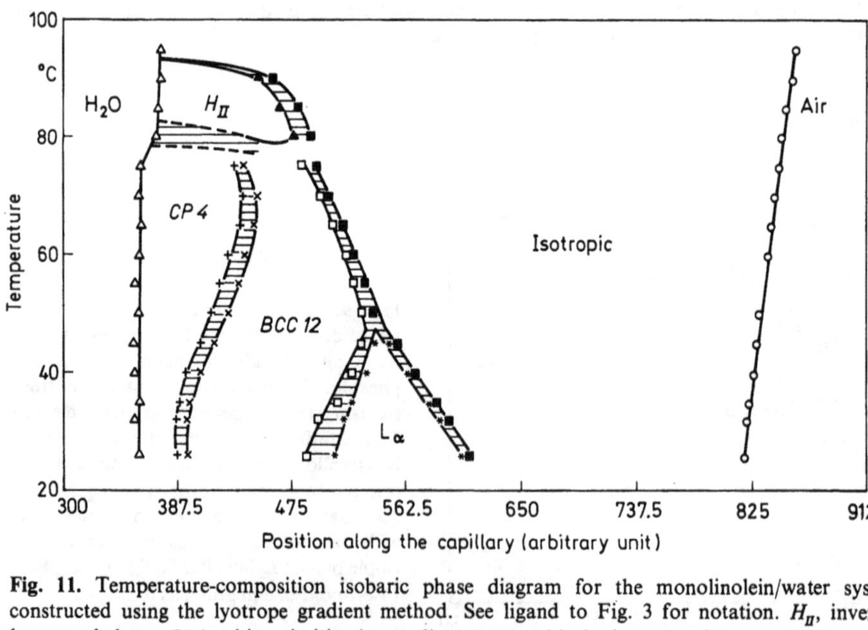

Fig. 11. Temperature-composition isobaric phase diagram for the monolinolein/water system constructed using the lyotrope gradient method. See ligand to Fig. 3 for notation. H_{II}, inverted hexagonal phase; *CP4*, cubic, primitive (aspect 8); *BCC12*, cubic, body-centered (aspect 12); Isotropic, fluid isotropic or melt

and in the lyotrope gradient method, to a particular lyotrope concentration. To collect phase information for the entire diagram, the procedure is repeated with samples representing different lines — either isopleths or isotherms.

Sample phase diagrams collected using the two gradient methods are presented in Fig. 9 and 11. It is important to realize that in both cases only the location of phase boundaries are shown, while in fact, phase identification and structural characterization is available *continuously* along each isotherm and isopleth in the respective plots.

3.2 Method Evaluation

The two new methods just described offer several advantages over conventional equilibrium methods:

- They facilitate rapid data collection by combining sample preparations that incorporate a range of conditions with a method of rapid phase identification and structural characterization.
- They are efficient. Because data collection is continuous along a temperature or composition gradient, the possibility that phases which exist in a narrow temperature or composition range will go undetected is eliminated.
- The gradients are flexible. The temperature range in the first method can be expanded or contracted by simply and rapidly adjusting the limiting temperatures on either end of the gradient rod. In the lyotrope gradient method, narrow phase fields can be extended by incubating the sample for a longer period of time or by adjusting the amount of lyotrope used.
- They utilize X-ray diffraction. X-ray diffraction allows *direct* qualitative and quantitative phase characterization — even in multiphase regions — and no potentially perturbing additives or molecular labels are needed. Although the high photon flux of synchrotron radiation is potentially damaging to the sample [15], particular parts need only be exposed to the beam for a short period of time and as a result, radiation damage is not a problem with this method.
- They have general applicability. Both methods apply to a wide range of materials including liquid crystals, polymers and nonaqueous solvents.

There are some limitations to these methods which must be addressed:

- They require access to a high intensity X-radiation source of the type available at a synchrotron radiation facility.
- They require a large amount of sample. However, this disadvantage is offset by the fact that both methods are nondestructive, and the samples can be recovered for reuse. Furthermore, the lyotrope gradient method in its present configuration requires only a single sample representing about 40 mg lipid to construct a complete phase diagram.
- Lyotrope gradient samples ideally should be fluid.
- Since both methods are based upon X-ray diffraction, difficulties are encountered in distinguishing liquid phases.
- In the lyotrope gradient method, lipid and lyotrope concentration along the length of the gradient must be established by independent means.

– Phase diagrams are meant to indicate phase behavior at equilibrium. In these methods, a non-equilibrium or steady-state approach is used to obtain equilibrium information. Although this theoretically is a point of contention, experimental evidence shows that phase diagrams created with these methods agree with and extend the information contained in phase diagrams produced with conventional equilibrium methods [20a, 23, 26].

4 Water Transport in Lyotropic Lipids

It was realized at an early stage in the development of the lyotrope gradient method described above that the transport properties of water in a contiguous series of lipid mesomorphs should be accessible. This provided the impetus for the development of a computer algorithm based on a finite-difference method for determining the mutual diffusion for water in the contiguous mesomorphs of the monoolein/water system [55a]. The FORTRAN program based on this algorithm searches for diffusion coefficients across a phase region such that calculated phase boundary positions coincide with moving boundary positions determined experimentally by TRXRD ([20a], refer to Sect. 3 for details). From the derived diffusion coefficients the phase development in time can be predicted. Another practical application is the prediction of the amount of time and lyotrope required to prepare samples in a defined mesomorphic state without mechanical mixing. In the case of excess water, about 60 days are needed to transform completely a 2.5 cm long monoolein sample initially in the (undercooled) fluid isotropic phase into the fully hydrated cubic primitive phase at 25 °C in the absence of mechanical mixing.

Results for the monoolein/water system give mutual diffusion coefficients ranging from $0.7\text{–}1.5 \times 10^{-6}$ cm/s in the fluid isotropic, lamellar liquid crystal and cubic phases ([55a], Fig. 7). These values are up to an order of magnitude lower than literature values of the self diffusion coefficients determined by pulsed field gradient NMR for this same system [96].

5 X-Ray Standing Waves: A New Molecular Yardstick for Model and Biomembranes

The discussion thus far has focused on bulk lipid systems. Our ultimate goal is to understand the behavior of lipids in cellular membranes which exist, in the main, as isolated or paired bilayers. Accordingly, there is a need for a method capable of reporting on the structural/positional details in isolated lipid mono-, bi- and trilayer systems as models for the biomembrane. As with most studies of thin layers, sensitivity is a problem due to the paucity of material sampled during measurement. In an effort to make good these deficits, my collaborators and I have implemented a new, long-period X-ray standing wave technique which gives structural information with sub-ångstrom resolution on thin Langmuir-Blodgett (LB) lipid films [7a].

The principle of the long-period X-ray standing wave method is as follows ([7a] and references therein). In the region of overlap between a coherently related incident and diffracted X-ray beam a standing wave is established both in and above the diffracting crystals as a result of interference. When used with perfect single crystals the conventional X-ray standing wave method is limited in its probing or sampling distance to the d-spacing of the diffracting crystal, typically 1 to 4 Å. The position of a heavy atom layer implanted in or deposited on the crystal relative to the diffracting planes of the crystal can be precisely determined by monitoring the X-ray fluorescence yield of the heavy atom as the crystal is rotated through the Bragg angle. In so doing, the antinodes of the X-ray standing wave shift inward by half a d-spacing during crystal rotation from just below to just above the Bragg peak. The fluorescence yield profile contains information on the location and distribution of the heavy atom layer. The E-field intensity is periodic along the stationary wave. Thus, by rocking the crystal through the Bragg angle the standing wave E-field sensed at the center of a heavy atom changes in a characteristic way and, through a photoelectric effect, gives rise to a characteristic fluorescence yield profile. Unfortunately, the modulo-d length scale of the method means that the conventional X-ray standing wave approach does not lend itself to studies of structures such as biomembranes which have a length scale ranging from 20–200 Å.

The X-ray standing wave method makes use of "fabricated crystals", so called layered synthetic microstructures (LSM), which can be prepared with diffracting plane spacings to match that of the material of interest. These are depth-periodic layered structures with alternating layers of electron dense and light atoms. Since the composition, surface layer and d-spacing of the LSM can be prepared to order, and since they offer reflectivities at the first-order Bragg peak as high as 80% and can support a well-defined standing wave, such materials are ideally suited for structural studies of biologically relevant membranes.

X-ray standing waves can be generated not only from interference between incident and diffracted beams but also between incident and specularly reflected (total external reflection) plane waves. The former probes with a sampling distance on the order of the d-spacing of the LSM. In contrast, the latter, so called long-period X-ray standing waves, has an inherently longer length scale and is useful in removing the modulo-d ambiguity from the first-order Bragg data.

In the present study, heavy atom soaps of arachidic acid (a C20 saturated fatty acid) were deposited as a trilamellar LB film on a tungsten/carbon LSM (d = 25 Å). Cadmium arachidate (CdA) was deposited as a marker layer in registry with the LSM surface followed by a bilayer of zinc arachidate (ZnA). The Z separation between the Zn and Cd layers and the coherent fraction (layer width) were tracked as a function of temperature in 10 °C increments from 35 °C to 140 °C. Initial measurements placed the Zn layer 53.4 Å above the LSM surface suggesting a slight tilt (approx. 17°) in the long axis of the chains relative to surface normal at low temperatures. Between 100 °C and 110 °C a dramatic shift in the Zn layer occurred which was shown by the specular standing wave data to correspond to an inward collapse by approx. 30 Å of the Zn layer position toward the LSM surface. No further change was observed upon continued heating or subsequent cooling of the sample. It is likely that the inward collapse corresponds, in part, to a chain order/disorder, tilting and/or interdigitation event since bulk, dry

CdA undergoes chain melting at ~120 °C [136a]. In-plane X-ray scattering measurements will help resolve this ambiguity.

These data demonstrate that the long-period X-ray standing wave method is useful as a molecular yardstick for establishing the arrangement of heavy atom layers in isolated model membranes with subångström resolution and for tracking heavy atom layer position through a thermotropic transition. The stage is now set to extend these measurements to studies on 1) heavy atom location in specifically and/or intrinsically labeled lipids, liquid crystals and proteins in deposited mono- and multilayers and 2) ion concentration profiles at model and reconstituted membrane/aqueous interfaces [7b].

Acknowledgements: Work included in this review was supported by NIH grants DK36849 (M. Caffrey), HL18255 (G. W. Feigenson), and RR01646 (J. K. Moffat), and NSF grant DMR81-12822 (B. W. Batterman). I am especially grateful to the following individuals who have contributed in their own unique ways to this review: D. Bilderback, G. Feigenson, H. Gerritsen, B. Hummel, J. Li, A. Mencke, D. Moynihan, W. Schildkamp, and M. Yeager. Special thanks go to S. Gruner, P. Laggner, B. Tenchov, and P. Westerman for providing unpublished data and to R. Magin for assistance with the microwave experiments.

6 References

1. Akiyama M (1981) Biochim. Biophys. Acta 644: 89
2. Akiyama M, Terayama Y, Matsushima N (1982) Biochim. Biophys. Acta 687: 337
3. Albert AD, Sen A, Yeagle PL (1984) Biochim. Biophys. Acta 771: 28
4. Amemiya Y, Wakabayashi K, Tanaka H, Ueno Y, Miyahara J (1987) Science 237: 164
5. Armitage D, Price FP (1978) J. Polym. Sci. Polym. Symp. 63: 95
6. Arndt UW (1986) J. Appl. Cryst. 19: 145
7. Batterman B, Ashcroft NW (1979) Science 206: 157

7a. Bedzyk MJ, Bilderback DH, Bommarito GM, Caffrey M, Schildkraut JS (1988) Science 241: 1788

7b. Bedzyk MJ, Bommarito GM, Caffrey M, Penner T (1988) CHESS Users Meeting, Ithaca, NY (Abstract)

8. Bilderback D, Moffat K, Owen J, Rubin B, Schildkamp W, Szebenyi M, Smith Temple B, Volz K, Whiting B (1988) Nucl. Instr. Meth. A266: 636
9. Bilderback DH, Moffat JK, Szebenyi DME (1984) Nucl. Instr. Meth. 222: 245
10. Blume A, Hillmann M (1986) Eur. Biophys. J. 13: 343
11. Blumenthal R (1987) Curr. Topics Membr. Transport 29: 203
12. Buras B, Materlik G (1986) Nucl. Instr. Meth. A246: 21
13. Bretcher MS (1987) Scientific American 257: 72
14. Brown GS, Lindau I (eds) (1986) Synchrotron Radiation Instrumentation Nucl. Instr. Meth. A246:
15. Caffrey M (1984) Nucl. Instr. Meth. 222: 329
16. Caffrey M (1985) Biochemistry 24: 4826
17. Caffrey M (1986) In: Leopold AC (ed) Membranes, metabolism and dry organisms. Comstock, Ithaca, New York p 242
18. Caffrey M (1987) Biochemistry 26: 6349
19. Caffrey M (1987) Biochim. Biophys. Acta 896: 123
20. Caffrey M (1987) Biophys. J. 51: 444a (Abstract)

20a. Caffrey M (1989) Biophys. J. 55: 47

21. Caffrey M, Bilderback DH (1983) Nucl. Instr. Meth. 208: 495

22. Caffrey M, Bilderback DH (1984) Biophys. J. 45: 627
23. Caffrey M, Bywater MT (1988) J. Soc. Cosmet. Chem. 39: 159
24. Caffrey M, Feigenson GW (1984) Biochemistry 23: 323
25. Caffrey M, Fonseca V, Leopold AC (1988) Plant Physiol. 86: 754
26. Caffrey M, Hing FS (1987) Biophys. J. 51: 37
27. Caffrey M, Lew RR (1986) Plant Cell Physiol. 27: 1091
28. Caffrey M, Mencke A (1988) CHESS Users Meeting, Cornell University, Ithaca, New York
29. Chang H, Epand RM (1983) Biochim. Biophys. Acta 728: 319
30. Chen SC, Sturtevant JM, Gaffney BJ (1980) Proc. Natl. Acad. Sci. U.S.A. 77: 5060
31. Chernavskii DS, Edius VL, Polezhaev AA (1981) BioSystems 13: 171
32. Cho KC, Choy CL, Young K (1981) Biochim. Biophys. Acta 663: 14
33. Clark NA, Rothschild KJ, Luippold DA, Simon BA (1980) Biophys. J. 31: 65
34. Clark R, Sigler P, Mills D (1988) Argonne National Laboratory. Workshop Report Number ANL/APS-TM-2
35. Corless JM, Costello MJ (1981) Exp. Eye Res. 32: 217
36. Crowe JH, Crowe LM, Carpenter JF, Rudolph AS, Wistrom CA, Spargo BJ, Anchordogny TJ (1988) Biochim. Biophys. Acta 947: 367
37. Cullis PR, DeKruijff B (1979) Biochim. Biophys. Acta 559: 399
38. Cullis PR, De Kruijff B, Hope MJ, Nayar R, Schmid SL (1980) Can. J. Biochem. 58: 1091
39. Cullis PR, DeKruijff B, Hope JM, Verkleij AJ, Nayar R, Tilcock CPS, Madden TD, Bally MB (1983) In: Aloia RC (ed) Membrane fluidity in biology Academic, New York, vol 1 p 39
40. Cunningham BA, Lis LJ, Quinn PJ (1986) Mol. Cryst. Liq. Cryst. 141: 361
41. Cunningham BA, Tamura-Lis W, Lis LJ, Quinn PJ (1988) J. Coll. Inter. Sci. 121: 193
42. Dafler JR (1977) J. Am. Oil Chem. Soc. 54: 249
43. Deamer DW, Leonard R, Tardieu A, Branton D (1970) Biochim. Biophys. Acta 219: 47
44. Demus D, Diele S, Grande S, Sackmann H (1983) Adv. Liq. Cryst. 6: 1
45. Dupont Y, Gabriel A, Chabre M, Gulik-Krzywicki T, Schechter E (1972) Nature 238: 331
46. Eck V, Holzwarth JF (1984) Surfactants in Solution 3: 2059
47. Eisenberger P (1986) Science 231: 687
48. Elamrani K, Blume A (1983) Biochemistry 22: 3305
49. Elamrani K, Blume A (1984) Biochim. Biophys. Acta 769: 578
50. Epand RM (1985) Chem. Phys. Lipids 36: 387
51. Frankel RD, Forsyth, JM (1979) Science 204: 622
52. Frankel RD, Forsyth JM (1985) Biophys. J. 47: 387
53. Franks F (1985) Biophysics and biochemistry at low temperatures, Cambridge University Press, New York
54. Genz A, Holzwarth JF (1986) Eur. Biophys. J. 13: 323
55. Genz A, Holzwarth JF, Tsong TY (1986) Biophys. J. 50: 1043
55a. Gerritsen H, Caffrey M (1989) J. Phys. Chem. (in press)
56. Gordon-Kamm WJ, Steponkus PL (1984) Proc. Natl. Acad. Sci. U.S.A. 81: 6373
57. Graham I, Gagne J, Silvius JR (1985) Biochemistry 24: 7123
58. Green RE (1977) Flash radiography symp. proc. American Society for Nondestructive Testing, Columbus, Ohio 43221
59. Grubb DT, Liu J-H, Caffrey M, Bilderback DH (1984) J. Polym. Sci., Polym. Phys. Ed. 22: 367
60. Gruenewald B (1982) Biochim. Biophys. Acta 687: 71
61. Gruenewald B, Blume A, Watanabe F (1980) Biochim. Biophys. Acta 597: 41
62. Gruenewald B, Frisch W, Holzarth, JF (1981) Biochim. Biophys. Acta 641: 311
63. Gruner SM (1987) Science 238: 305
64. Gruner SM (1985) Proc. Natl. Acad. Sci. U.S.A. 82: 3665
65. Gruner SM, Milch JR, Reynolds GT (1982) Nucl. Instr. and Meth. 195: 67
66. Hammes GG, Tallman DE (1970) J. Am. Chem. Soc. 92: 6042
67. Helliwell JR (1984) Rep. Prog. Phys. 47: 1403
68. Holzwarth JF, Frisch W, Gruenewald B (1982) Microemulsions In: Robb ID (ed) Proc. Conf. Phys. Chem. Microemulsions, Plenum, New York p 185
69. Hui SW, Stewart TP, Boni LT (1981) Science 212: 921
70. Hui SW, Stewart TP, Yeagle PL (1980) Biochim. Biophys. Acta 601: 271
71. Hyde ST, Andersson S, Ericsson B, Larsson K (1984) Z. Krist. 168: 213

72. Hyde ST, Andersson S, Larsson K (1986) Z. Krist. 174: 237
73. Inoue S, Nishimura M, Yasunaga T (1981) J. Phys. Chem. 85: 1401
74. Inoue T, Ohshima H, Kamaya H, Ueda I (1985) Biochim. Biophys. Acta 818: 117
75. Jabarin SA, Stein RS (1973) J. Phys. Chem. 77: 409
76. Jain MK, Wu NY-M, Wray LV (1975) Nature, 255: 494
77. Janesick J, Blouke M (1987) Sky and Telescope, 74: 238
78. Jensen JW, Schutzbach JS (1984) Biochemistry 23: 1115
79. Johnson ML, Van Osdol WW, Frasier SG (1985) Comments Mol. Cell. Biophys. 3: 77
80. Johnson ML, Winter TC, Biltonen RL (1983) Anal. Biochem. 128: 1
81. Johnson ML, Van Osdol WW, Biltonen RL (1986) Methods Enzymol. 130: 534
82. Kalata K (1981) IEEE Trans. Nucl. Sci. NS-28: 852
86. Koynova RO, Boyanov AI, Tenchov BG (1987) Biochim. Biophys. Acta 903: 186
87. Krog NJ, Riisom TH, Larsson K (1985) In: Bercher P (ed) Encyclopedia of emulsion technology: applications. Marcel Dekker, New York, vol 2 p 321
88. Laggner P (1987) Springer Ser. Biophys. 1: 281
89. Laggner P (1988) Topics in Current Chemistry 145: 173
89a. Laggner P, Kriechbaum M, Hermetter A, Paltauf F, Hendrix J, Rapp G (1989) Trends Coll. Interface Sci. (in press)
90. Laggner P, Lohner K, Muller K (1987) Mol. Cryst. Liq. Cryst. 151: 373
91. Laggner P, Muller K, Lipka G, Lohner K (1985) Int. Surf. Coll. Sci. Meet, Potsdam, NY.
92. Larson BC, White CW, Noggle TS, Barhorst JF, Mills DM (1983) Appl. Phys. Lett. 42: 282
92a. Laughlin RG (1984) In: Tadros TF (ed) Surfactants, Academic, London, p 53
93. Lentz BR, Freire E, Biltonen RL (1978) Biochemistry 17: 4475
94. Leopold AC (1986) Membranes, metabolism and dry organism, Comstock, Ithaka, NY, p 374
95. Levine YK (1973) Progr. Surf. Sci. 3: 279
96. Lindblom G, Larsson K, Johansson L, Fontell K, Forsen S (1979) J. Am. Chem. Soc. 101: 5465
97. Lis LJ, Quinn PJ (1986) Biochim. Biophys. Acta 862: 81
98. Lis LJ, Quinn PJ (1986) Mol. Cryst. Liq. Cryst. 140: 319
99. Lis LJ, Quinn PJ (1987) Mol. Cryst. Liq. Crys. 146: 35
100. Longley W, McIntosh T (1983) Nature 303: 612
101. Lutton ES (1965) J. Am. Oil Chem. Soc. 42: 1068
102. Luzzati V (1968) In: Chapman D (ed) Biological membranes, physical fact and function. Academic, New York, vol 1, p 71
103. Mariani P, Luzzati V, Delacroix H (1988) J. Mol. Biol. 204: 165
104. Mackay AL (1985) Nature 314: 604
104a. Mateu L, Kirchhausen T (1979) Acta Cient. Venez. 30: 478
105. Mattai J, Shipley GG (1986) Biochim. Biophys. Acta 859: 257
106. Mayorga OL, Lacomba JL, Freire E (1988) Biophys. J. 53: 125a (Abstract)
107. McCray J, Trentham DR (1988) Ann. Rev. Biophys. Chem. 18: 239
108. Melchior DL, Bruggemann EP, Stein JM (1982) Biochim. Biophys. Acta 690: 81
109. Miller IR, Bach D (1986) Biochim. Biophys. Acta 863: 121
110. Moffat K (1988) Ann. Rev. Biophys. Chem. 18: 309
111. Mollevanger LCPJ, DeGrip WJ (1984) FEBS Lett. 169: 256
112. Nagle JF, Wilkinson DA (1982) Biochemistry 21: 3817
113. Navarro J, Toivio-Kinnucan M, Racker E (1984) Biochemistry 23: 130
114. Noordam PC, Van Echteld CJA, De Kruijff B, DeGier J (1981) Biochim. Biophys. Acta 646: 483
115. Owen JD, Hemmes P, Eyring EM (1970) Biochim. Biophys. Acta 219: 276
116. Patton JS, Carey MC (1979) Science 204: 145
117. Persson PKT (1984) Chem. Phys. Lipids 35: 11
117a. Phonphok N, Westerman PW, Lis LJ, Quinn PJ (1989) J. Coll. Interfac. Sci. 127: 487
118. Price FP, Wendorff JH (1971) J. Phys. Chem. 75: 2839
119. Price FP, Wendorff JH (1971) J. Phys. Chem. 75: 2849
120. Pringle MJ, Chapman D (1980) Biochem. Soc. Trans. 8: 686
121. Quinn PJ, Lis LJ (1986) Biochem. Soc. Trans. 14: 650
122. Quinn PJ, Lis LJ (1987) J. Coll. Interfac. Sci. 115: 220
123. Ranck JL (1983) Chem. Phys. Lipids 32: 251

124. Ranck JL, Lattelier L, Shechter E, Krop B, Pernot P, Tardieu A (1984) Biochemistry 23: 4955
125. Rand RP, Sengupta S (1972) Biochim. Biophys. Acta 255: 484
126. Rega H-U, Sackmann E (1974) Ber. Bunseng. 78: 915
127. Rowe EM (1981) Physics Today 34: 38
128. Rowe EM, Weaver JH (1977) Sci. Am. 236: 32
129. Ruocco MJ, Shipley GG (1982) Biochim. Biophys. Acta 691: 309
130. Serrallach EN, Dijkman R, deHaas GH, Shipley GG (1983) J. Mol. Biol. 170: 155
131. Shipley GG (1973) in: Chapman D, Wallach DFH (eds) Biological Membranes, Academic, New York, vol 2 p 1
132. Shyamsunder E, Gruner SM, Tate MW, Turner DC, So PTC, Tilcock CPS (1988) Biochemistry 27: 2332
133. Siegel DP (1984) Biophys. J. 45: 399
134. Siegel DP (1986) Biophys. J. 49: 1171
135. Siegel DP (1987) Chem. Phys. Lipids 42: 279
136. Small DM (1987) Handbook of lipid research: The physical chemistry of lipids from alkanes to phospholipids. Vol 4. Plenum, NY 672 p
136a. Spegt PA, Skoulios AE (1963) Acta Cryst. 16: 301
137. Steponkus PL (1984) Ann. Rev. Plant Physiol. 35: 543
138. Strehlow U, Jahnig F (1981)Biochim. Biophys. Acta 641: 301
139. Subezynski WK, Kusumi A (1986) Biochim. Biophys. Acta 854: 318
140. Tate MW (1987) Equilibrium and kinetic states of the Lα-HII phase transition. Ph.D. thesis, Princeton Univ., Princeton
141. Tenchov BG (1986) Chem. Phys. Lipids 39: 155
142. Tenchov BG, Lis LJ, Quinn PJ (1987) Biochim. Biophys. Acta 897: 143
142a. Tenchov BG, Lis LJ, Quinn PJ (1988) Biochim. Biophys. Acta 942: 305
143. Tessie J (1979) Biochim. Biophys. Acta 555: 553
144. Tilcock CPS, Bally MB, Farren SB, Cullis PR, Gruner SM (1984) Biochemistry 23: 2696
145. Trundell JR (1977) Anesthesiology 46: 5
146. Tsong T-Y (1974) Proc. Natl. Acad. Sci. U.S.A. 71: 2684
147. Tsong T-Y, Kanehisa MI (1977) Biochemistry 16: 2674
148. Tsuchida K, Hatta I, Imaizumi S, Ohki K, Nozawa Y (1985) Biochim. Biophys. Acta 812: 249
149. Tummler B, Hermann U, Maass G, Eibl H (1984) Biochemistry 23: 4068
150. Van Echtfeld CJA, Van Stigt R, DeKruijff B, Leunissen-Bijvelt J, Verklkeij AJ, DeGier J (1981) Biochim. Biophys. Acta 648: 287
151. Van Venetic R, Verkleij AJ (1981) Biochim. Biophys. Acta 645: 262
152. Whiting BR, Owen JF, Rubin BH (1988) Nucl. Instr. Meth. A266: 628
153. Winick H, Brown G, Halbach K, Harris J (1981) Physics Today 34: 50
154. Winick H, Doniach S (1980) Synchrotron Radiation Research, Plenum, New York
155. Wu W-G, Chong PL-G, Huang C-h (1985) Biophys. J. 47: 237
156. Wu W-G, Huang C-h (1983) Biochemistry 22: 5068
157. Wu W-G, Huang C-h, Conley TG, Martin RB, Levin IW (1982) Biochemistry 21: 5957
158. Yager P, Peticolas WL (1982) Biochim. Biophys. Acta 688: 775
159. Yoshino H, Kobayashi M, Samejima M (1982) Chem. Pharm. Bull. 30: 2941
160. Yoshino H, Kobayashi M, Samejima M (1982) Chem. Pharm. Bull. 29: 2661

Research on Synthetic Polymers by Means of Experimental Techniques Employing Synchrotron Radiation

Rainer Gehrke

Hamburger Synchrotronstrahlungslabor (HASYLAB)/DESY Notkestrasse 85, D-2000 Hamburg 52, FRG

Table of Contents

1 Introduction

This article is intended to give a survey of the application of synchrotron radiation (S.R.) in polymer science. Although several experimental techniques exist in the field of polymer physics and chemistry which could exploit the advantages of S.R., the main effort lies on the various X-ray scattering methods. Especially numerous time resolved small-angle X-ray scattering measurements facilitated by S.R. for the first time and many questions that have been answered by these experiments.

The first investigations using S.R. in polymer research were performed about ten years ago. Since that time a perpetually growing number of experiments has been done year by year. In meantime it has become difficult to to keep up with all the work that has been accomplished in this field. The aim of the present article is to light up the different topics of polymer science in which S.R. is utilized. In order to do this, typical examples for each of these topics are selected. Some of them are already to be found in a former review [1], but a number of more recent results have been added.

At the beginning a brief summary of the basic principles of X-ray scattering and scattering instrumentation will be given. These remarks will be very closely related to the specific peculiarities of polymer research and to the experiments described later. Items, which do not fall into this narrow category are mostly omitted. Here the reader is referred to more detailed literature.

2 X-Ray Scattering in Polymers

2.1 Basic Principles of X-Ray Scattering in Polymers

Regarding the elastic scattering of X-rays of a given wavelength, then from the experimental point of view, two methods can be distinguished. In wide angle X-ray scattering (WAXS) the intensity at scattering angles above $2\theta \approx 2.5°$ is registered, while a small angle X-ray scattering (SAXS) experiment measures the intensities scattered at lower angles. In the latter case special instrumental provisions have to be made in order to separate the scattered intensity sufficiently from the extremely intensive primary beam. Usually this is done by collimating the primary beam as perfectly as possible using sets of narrow slits and by providing large distances between the sample and detection plane.

Generally, X-ray scattering reflects periodical fluctuations of the electron density within the sample [2, 3, 4]. At a typical wavelength of 0.15 nm, which corresponds to the widely used CuK_α radiation, WAXS is normally caused by intra- and intermolecular distances. In liquid or amorphous polymer materials the WAXS is determined by the distance distribution of the chain segments of the macromolecules, while in crystalline-ordered polymers the maxima of the WAXS reflections are determined by the distances of certain netplanes. The angular positions 2θ of the reflection maxima and the distances d of the corresponding structures are related to each other by the well known Bragg equation,

$$n\lambda = 2d \sin \theta \, , \tag{1}$$

where n is the order of the reflection and λ is the wavelength of the utilized radiation.

If there are different domains leading to distinguishable WAXS intensity profiles, e.g. crystalline and amorphous phases coexisting in separate regions of a sample, the amount of each phase can be determined simply by separating the integral intensities which contribute to the WAXS from the different phases. In the case of crystalline and amorphous domains this analysis will lead to the degree of crystallinity.

A further evaluation of WAXS data can give information about the degree of lattice distortions in the crystalline regions [2, 4], which lead to a broadening of the reflections and a diffuse background scattering. In disordered systems the correlation function of the molecular distances can be obtained.

The SAXS has its origin in periodical fluctuations of higher wavelengths compared to those given by intermolecular distances. They are caused by molecular superstructures, which are more or less regularly arranged relative to each other. These superstructures can be crystalline ordered regions which are embedded in an amorphous surrounding or can be domains of one sort of material within a matrix of another material. Typical dimensions accessible by SAXS measurements lie in the range from a few up to some hundred nanometers. In the case of a sufficiently regular arrangement of the scattering domains reflection maxima appear in the scattering profile. In that case again the Bragg equation (1) is valid, where d is now given by the mean distance between adjacent domains of the same type.

Another important relation is valid for the integral SAXS intensity, the scattering power Q. In the case of a two phase system its value is related to the volume fraction x of one of the two domains by [3]

$$Q = \int I(q)\, d^3q = CVx(1 - x)\,(\Delta\varrho)^2 \tag{2}$$

$\Delta\varrho$ is the electron density difference between the two different kinds of domains, V is the irradiated volume, and C is an instrumental constant, which depends on the geometry of the scattering experiment and on the primary beam intensity.

A more detailed analysis of the SAXS intensity profile leads to conclusions about shape and size distribution of the domains [3, 5–10] and about the width and nature of the interfacial layers bounding the domains [11–14]. If there is a long range order between the domains, information about the degree of this order can be obtained.

Numerous macromolecular materials are known to be partially crystalline [15]. The crystalline regions are very often lamellar shaped and are formed by backfolding of the chains, as it is schematically drawn in Fig. 1a. The lamellae build up stacks, in which two adjacent lamellae are separated by an amorphous layer of disordered chains. The stacks might form further superstructures of still larger dimensions. For example, in the case of the crystallization of an unoriented amorphous material, the crystallization process often starts from nucleation centers. Then, from each center a radial symmetric structure, called a spherulite, proceeds to

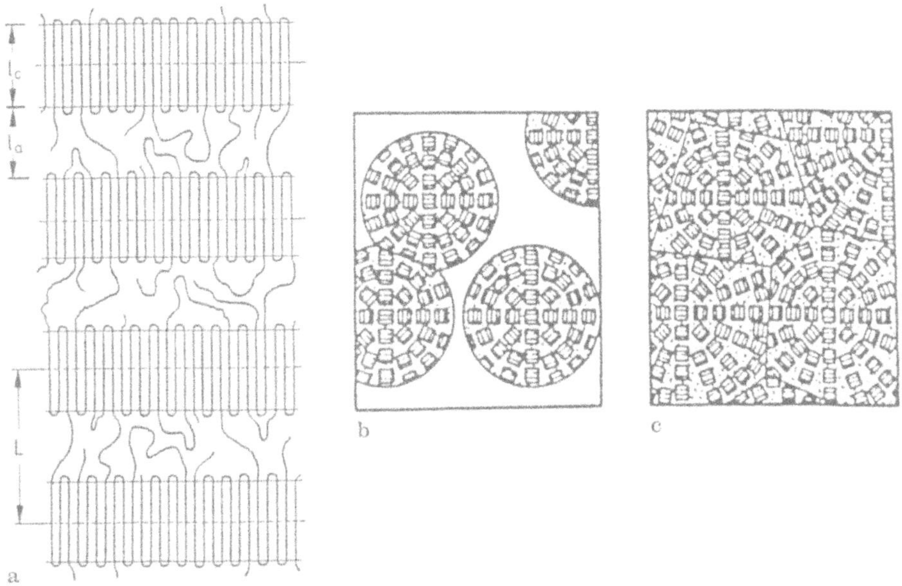

Fig. 1a–c. Schematic representation of typical morphology in partially crystalline polymers; **(a)** stack of crystalline lamellae of thickness l_c and amorphous regions of thickness l_a (L = long period); **(b)** spherulites formed by a radial symmetric arrangement of stacked lamellae, during main crystallization; **(c)** spherulites after the end of main crystallization

grow. These spherulites are formed by stacks of lamellae which are all oriented with their surfaces perpendicular to the spherulite radius (see Fig. 1b). During the main crystallization the spherulites grow, until they join each other and fill the whole volume of the sample (see Fig. 1c). After this main crystallization often an additional increase of the degree of crystallinity is observed. During this residual crystallization, additional growing of the crystal lamellae and improvement of the crystal perfection takes place.

In a partially crystalline polymer, WAXS is a superposition of the diffuse scattering from the amorphous regions (amorphous halo) and the crystal reflections from the crystalline regions. The latter show normally a significant broadening due to lattice distortions. Figure 2a gives, as an example, the WAXS of polyethylene terephthalate with a degree of crystallinity of about 50%. The crystals form a triclinic structure, the miller indices of the different reflections are also given in the figure.

The SAXS of the same material is shown in Fig. 2b. The observed maximum is due to the regular arrangement of crystal lamellae and amorphous regions in a periodical order, as described above. Therefore with Eq. (1), one can calculate the mean distance of the centers of gravity of two adjacent lamellae. This value is often termed long period. The expression for the scattering power Q, Eq. (2) can be rewritten as

$$Q = CVx_s w_{cs}(1 - w_{cs})(\varrho_c - \varrho_a)^2 \qquad (3)$$

where x_s is the volume fraction of the sample that is filled with spherulites (i.e. $x_s = 1$ at the end of the main crystallization) and w_{cs} is the volume fraction of a single

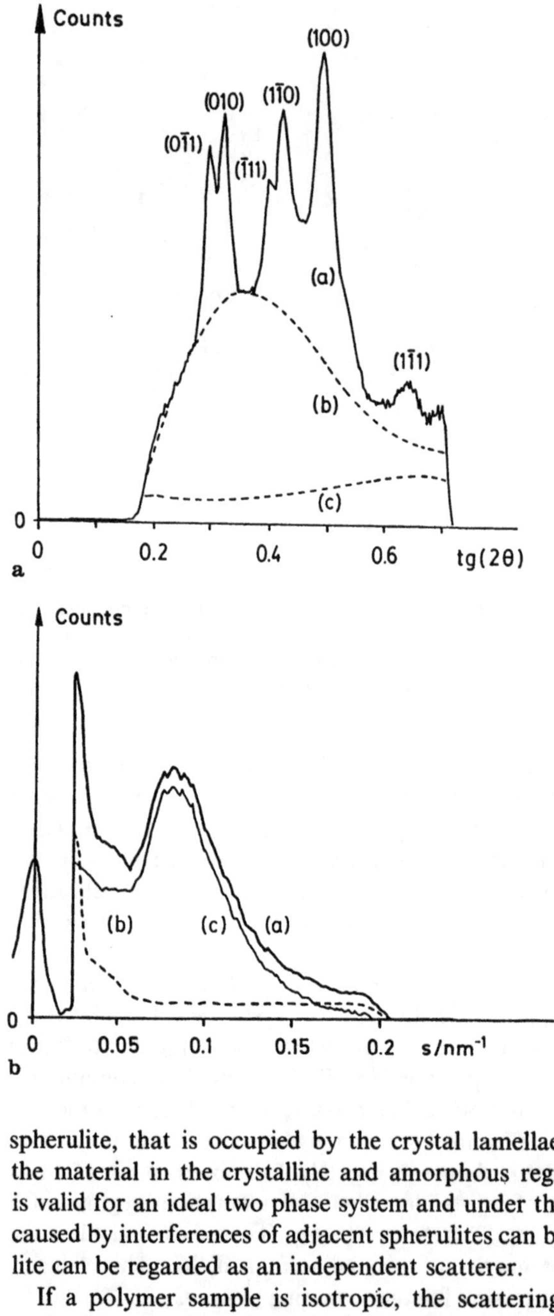

Fig. 2a, b. Wide-angle X-ray scattering of well crystallized polyethylene terephthalate; **(a)** total scattering profile; **(b)** fitted scattering curve of a completely amorphous sample; **(c)** background scattering **2b.** Small-angle X-ray scattering of well crystallized polyethylene terephthalate; **(a)** total scattering profile; **(b)** background scattering; **(c)** curve (a) corrected for background scattering

spherulite, that is occupied by the crystal lamellae. ϱ_c and ϱ_a are the densities of the material in the crystalline and amorphous regions, respectively. This equation is valid for an ideal two phase system and under the assumption, that the scattering caused by interferences of adjacent spherulites can be neglected, i.e. that each spherulite can be regarded as an independent scatterer.

If a polymer sample is isotropic, the scattering intensity at a given scattering angle does not depend on the orientation of the sample in the laboratory frame. But if the sample has been drawn before or after the crystallization, the molecules and consequently the crystalline regions become oriented. This has its manifestation in WAXS and SAXS. The scattered intensity becomes anisotropic, i.e. it depends on the orientation of the sample relative to the geometry of the scattering experiment.

Measurements of this anisotropy lead to conclusions about the nature and degree of the orientation of the crystals and of the chain molecules themselves [16].

In general, polymer materials are weak X-ray scatterers, because they mainly consist of light elements (H, C, N, O). Additionally, in the case of SAXS, the electron density differences between the various domains are normally rather small, resulting in a weak scattering power according to Eq. (2) or Eq. (3). In the case of WAXS the large volumes of the crystallographic unit cells usually found in polymer crystals and the strong broadening of the reflections due to lattice distortions imply low intensities to be scattered to the different angles.

For this reasons, the recording of a scattering diagram with a sufficient signal to noise ratio using conventional X-ray sources is rather time consuming. A SAXS pattern of a thin polymer film might require exposure times of several days, even if position sensitive detectors are used. Time resolved measurements to follow relatively fast structural changes taking place in a time range from fractions of a second to several minutes, are obviously impossible to perform under these conditions.

In this context the high photon flux of S.R. sources has opened impressive new possibilities. In addition, with conventional sources a large number of the produced photons are rejected by the collimating system. This becomes an especially serious limitation in the case of SAXS, where extremely fine collimation is necessary. Therefore not only the high flux, but also the high brightness of the S.R. sources is of great importance, S.R. is already well collimated at the moment of its production. Taking this high brightness into account, SAXS equipment installed at a bending magnet of a 3.5 GeV storage ring, operating at 100 mA and supplied with a perfect crystal monochromator will produce scattering intensities which are more than 2000 times higher than with a 2.4 kW rotating anode X-ray tube, utilizing CuK_α radiation. The reader can imagine the drastical reduction of exposure times.

For this reason, at several S.R. facilities, instruments have been designed which can be used excellently for X-ray scattering experiments with polymers. The most important beamlines exist at the storage ring DORIS at DESY in Hamburg (FRG), at the SRS in Daresbury (GB), at the SSRL in Stanford (USA), at the CHESS in Cornell (USA), at the NSLS in Brookhaven (USA), at L.U.R.E. in Orsay (F), and at the Photon Factory in Tsukuba (JPN).

2.2 X-Ray Scattering Instruments at Synchrotron Radiation Beamlines

2.2.1 Focussing Instruments

Focussing X-ray scattering instruments are widely used for time resolved WAXS and SAXS experiments in combination with position sensitive detectors. As an example for this type of equipment a brief description of an instrument will be given, which has been built at the Hamburg Synchrotron Radiation Laboratory (HASYLAB) at the storage ring DORIS at DESY. This instrument is dedicated especially to polymer research. Figure 3 shows schematically its design. The S.R. produced by a bending magnet comes from the left side. The size of the source is about 2×3 mm^2. The primary beam first impinges on a flat triangular shaped perfect germanium crystal. A monochromatization takes place at the (111) netplanes of the crystal, the instrument only makes use of a narrow band of wavelengths around $\lambda = 0.15$ nm and

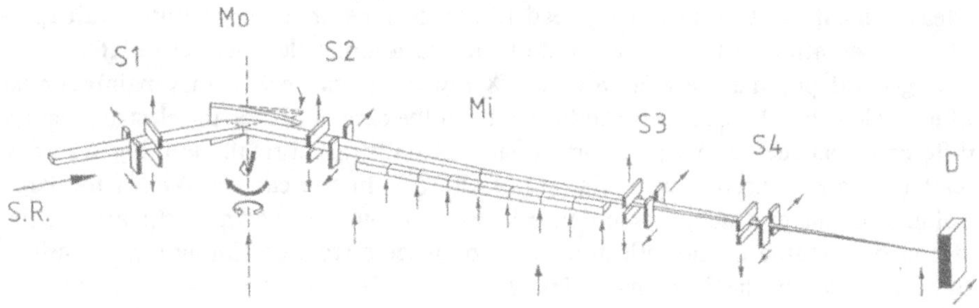

Fig. 3. Schematic representation of the SAXS/WAXS Polymer Beamline A2 at HASYLAB/Hamburg; *S1*, *S2*: slits, *Mo*: crystal monochromator, *Mi*: quartz mirror, *S3*: aperture slit, *S4*: guard slit and sample position, *D*: detector

of width $\Delta\lambda/\lambda = 10^{-4}$. The crystal can be bent by applying an adjustable force to the vertex of the triangle. This bending leads to a focussing of the beam in the horizontal direction. From the monochromator the beam is falling under grazing incidence onto a mirror consisting of eight flat quartz plates. The plates can be adjusted relative to each other in order to approach to an elliptical surface. By this one gets a focussing in the vertical direction. Additionally the mirror reduces the unwanted third-order radiation, which also has passed the monochromator. The primary beam is focussed onto the detection plane, where a demagnified image of the source is obtained. The distance from source to mirror comes to 20 m and the detection plane is 5 m from the mirror. The sample can be situated up to 3.5 m in front of the detector, depending on the range of scattering angles to register. Close behind the mirror is mounted an aperture slit, a second slit is situated in front of the sample. This guard slit rejects the stray light produced at the aperture slit. The resolution, which is theoretically attainable with this design, i.e. the largest periodical distance that can be measured, can be calculated from [1]

$$L = \frac{\lambda}{2}\left\{\frac{1}{2}\sin\left(\tan^{-1}\frac{m}{L_2}\right)\right\}^{-1} \tag{4}$$

with

$$m = \frac{S_1}{2L_a}(L_b + L_2), \qquad L_a = \frac{S_1}{S_2}L_b,$$

and

$$L_b = \frac{L_1 S_2}{S_1 + S_2}.$$

In this formula L_1 is the distance aperture slit — guard slit, S_1 the width of the aperture slit, L_2 the distance guard slit — detection plane, and S_2 is the width of the guard slit which can be calculated from

$$S_2 = \frac{S_1 - w}{L_1 + L_2}L_2 + w, \tag{4a}$$

with w being the size of the primary beam in the detection plane. For the instrument described above the maximum resolution is calculated to be about 100 nm.

Concerning time resolved measurements, special care must be taken of the problem how to monitor the time course of the primary beam intensity during the experiment. Another problem is, how to observe changes in the absorption of X-rays within the sample, when structural changes occur. The measured scattering patterns have to be corrected according to both effects.

The simplest solution of the first problem is to place an ionization chamber somewhere between aperture slit and sample. Another possibility is to place there a thin Kapton foil, inclined 45° to the primary beam. The small portion of intensity, that is scattered by the foil, can then be registered by a scintillation counter. If a second foil is mounted close behind the sample, then the absorption of X-rays in the sample can be measured additionally by comparing the intensities registered at the two foils. This design has been realized at some SSRL beamlines in Stanford. At the polymer beamline at the HASYLAB it became evident, that fluctuations of the primary beam position can lead to fluctuations of the intensity at the sample, which cannot be registered by a monitor placed in front of the sample. This observation is obviously caused by restrictions of the beam directly at the sample holders.

An attempt to circumvent this problem is to register a small portion of the primary beam instensity on the detector, together with the scattering pattern. The primary beam stop, a small block of lead mounted in front of the detector in order to prevent the primary beam from passing onto it, was provided with a central borehole. This hole was filled with aluminum of appropriate thickness to diminish the primary beam. Thus a weak portion of it was allowed to pass to the detector. Now all measurements could be conveniently corrected to the same integral intensity. However, care must be taken, if this semitransparent beam stop is also used to correct absorption effects. The aluminum layers are usually several millimetres thick and therefore absorb entirely the radiation of the first order harmonic coming from the monochromator. This indicates, that the observed photons belong exclusively to third order radiation. The semitransparent beamstop therefore is sensitive to changes in the absorption of these photons and not to the absorption of the first order ones, the latter will be markedly higher [35].

An elegant solution is to integrate an ionization chamber directly into the primary beam stop. This involves several technical problems, but has been tried already successfully at the HASYLAB.

2.2.2 High Resolution Instruments

Structures with dimensions above a few hundred nanometers cannot be resolved by any of the existing focussing cameras. An instrument with a significantly higher resolution is provided by the Bonse Hart type camera [17]. It does not contain any focussing optical elements. The X-ray beam is collimated and monochromatized to an extremely high extent by multiple Bragg reflection at perfect crystals. Figure 4 shows schematically an instrument constructed by Bonse et al. at HASYLAB [18]. The primary beam is first monochromatized by a double crystal monochromator. Then a collimation in the horizontal direction is performed by a multiple reflecting grooved crystal. This is a monolithic channel cut germanium crystal, which collimates

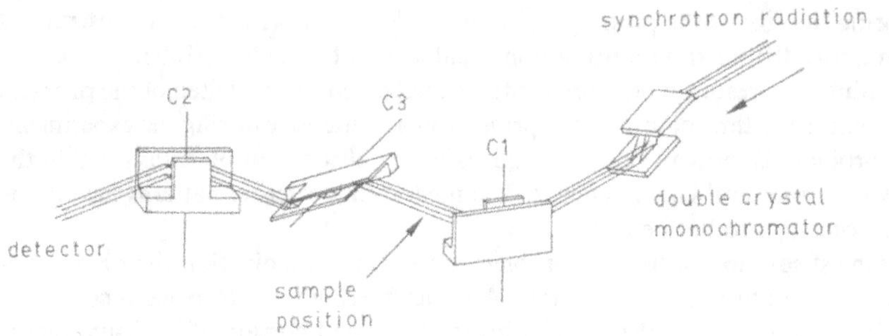

Fig. 4. Schematic representation of the ultra small angle scattering camera at HASYLAB/Hamburg

the beam by three consecutive (220) reflections. After passing the crystal, the beam falls onto the sample. The scattered radiation is then collimated in two directions by two other channel cut crystals, which are identical to the first one and are oriented perpendicular to each other. Only scattered photons which correspond to a well defined momentum transfer are allowed to pass this analyser arrangement and can be detected. The scattering curves are obtained by rotating the horizontal or vertical analyzer crystal about an axis perpendicular to the diffraction plane, as signed as C2 or C3 in Fig. 4. With this design correlation lengths up to 2.8 μm can be resolved. With silicon instead of germanium crystals, the resolution can be increased to 6.5 μm. Since the scattering angle has to be scanned, time resolved measurements cannot be performed.

This instrument favourably utilizes S.R. as an X-ray source of high brightness, because due to the high degree of collimation only a small fraction of a divergent photon beam can pass to the detector.

A feasibility test of the described camera was recently performed with different polystyrene latex samples forming scattering particles of various sizes [19]. The results can be seen in Fig. 5.

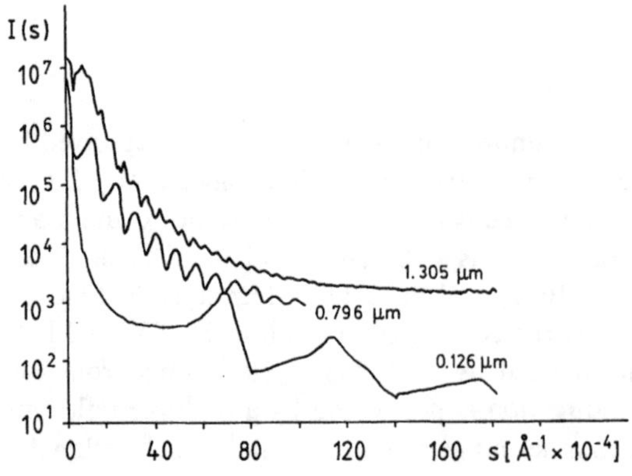

Fig. 5. Scattering curves of styrene latex samples of various indicated particle diameters obtained with the ultra small angle scattering camera of Fig. 4 [19]

2.2.3 Engergy Dispersive Instruments

With its broad spectrum of wavelengths, S.R. offers the possibility of performing energy dispersive scattering experiments [20]. Instead of measuring the intensity of the scattered radiation at a fixed wavelength and as a function of the scattering angle, in an engergy dispersive experiment the intensity at a fixed angle is observed as a function of the wavelength or energy. The incident beam must contain a continuous spectrum and the detector must be energy dispersive with a sufficient energy resolution. The main advantage of this technique is the optimal exploitation of the photons offered by the source, since not only a narrow energy band is utilized as in the case of angular dispersive measurements. Another advantage is that the scattered radiation is only registered in a small interval of scattering angles. This might simplify the construction of windows in sample holders, ovens, high pressure cells etc. under critical experimental conditions, where only small exit windows are available.

In spite of these benefits, energy dispersive measurements with polymer samples have been performed only in quite a few cases [21]. The main objections against this method are, that the high photon flux through the sample could seriously damage the investigated materials, and that presently no detectors exist which are fast enough to perform time resolved measurements.

2.2.4 Instruments with Variable Wavelength

Some X-ray scattering instruments at S.R. sources provide the option to change the wavelength of the primary beam. A typical setup in connection with focussing instrumentation consists of a double crystal monochromator designed with two flat single crystals in a non-dispersive arrangement. The focussing is then realized with bent mirrors. An example is the X-20C beamline at the NSLS in Brookhaven [78], where a toroidal mirror is placed in front of the double crystal monochromator. The 1:1 optics produce a $1 \times 1\ mm^2$ focal spot at the detection plane, the available photon energies lie in the range from 5 keV to 12 keV.

The scattering pattern of a given structure shifts gradually towards higher scattering angles, when the wavelength is increased. Therefore, changing to a higher wavelength, in a SAXS experiment can mean to register at more convenient scattering angles or equivalently to increase the resolution, i.e. structures of larger size can be measured.

On the other hand, increasing of the wavelength implies also increasing of the absorption of the X-rays within the sample. For this reason the wavelength corresponding to the CuK_α radiation (i.e. $\lambda = 0.15$ nm) is often choosen as a good compromise when investigating polymer samples.

Another possibility offered by a variable wavelength is the measurement of the scattering patterns at different wavelengths above and below the absorption edge of a certain element contained in the sample material. The anomalous scattering technique can supply enhanced information about the spatial arrangement of these special atoms. However, the typical elements contained in polymers have low atomic numbers, thus the absorption edges are at low energies. Stuhrmann [22] has designed an instrument for the HASYLAB in Hamburg which has a variable wavelength of between 0.12 nm and 0.6 nm and thus covers the K-absorption edges down to phosphorus. Scattering angles of up to 60° can be observed. However, measurements in the low energy region demand very thin samples to overcome the absorption problems.

2.3 X-Ray Detectors for Position Sensitive Measurements

Time resolved measurements of X-ray scattering usually require the application of position sensitive detectors with electronical readout schemes. They allow the simultaneous registration of the scattering pattern along the whole range of interesting angles. This leads to a significant reduction of the data acquisition times and is a general supposition for performing dynamical investigations.

The mainly demanded properties of these detectors are high sensitivity, a wide and linear dynamic range, and the ability to work with high photon rates. This requirements take into account, that especially with SAXS vast intensity differences occur and in the case of fast time resolved measurements, great amounts of photons have to be registered in relatively short time intervals.

The following chapters will give some remarks on those types of detectors, which are presently used in X-ray scattering experiments with polymers.

2.3.1 Gas Filled Detectors

The widely used one-dimensional position sensitive proportional counters utilize the delay-line principle [23]. A thin anode wire is tightened in front of a cathode which consists of a number of parallel metal strips connected to each other by an electrical delay-line. This arrangement is enclosed by a gas filled case. A beryllium window allows the X-ray photons to enter into the gas filled area. The photons ionize the gas and thus produce positive ions, which are collected by the cathode. This causes an electrical pulse at the cathode, which immediately starts to migrate towards each end of the delay-line. From the time difference between the appearance of the signal at both ends of the line, one can localize the point at which the photon has ionized the gas. A typical available cathode length is 80 mm, the spatial resolution being 250 μm. The dead time after the occurrence of a photon event is about 0.8 μsec. S.R. sources emit the photons in short light pulses, due to the bunch structure of the particle beams in the storage rings. Since the duration of a single pulse is very short compared to the dead time of the counter, it cannot resolve more than one scattered photon per bunch. This implies, that with strongly scattering samples, the primary beam often has to be reduced in order to provide an optimal reduction of the number of bunches, during which more than one photon is scattered into the detector. From statistical arguments it follows, that with a bunch rate of 10^6 sec^{-1} the maximum counting rate at the detector should not significantly exceed 5×10^4 sec^{-1} [24]. This limit is confirmed by practical experience.

If the scattering patterns are anisotropic due to a preferred orientation of molecules and crystals, it is often desirable to use two-dimensional position sensitive counters. Such area detectors, based on the principles described above, are the multiwire proportional counters (MWPC). A number of parallel tightened anode wires are mounted in front of a cathode structure, the latter consisting of two layers of parallel metal strips which are oriented horizontally in one layer and vertically in the other. One possibility of reading out the strips is to connect the strips of each layer to an independent delay-line. Then the signals from these two delay-lines give the vertical and horizontal coordinates of the photon event. The electronical expense of this design is relatively low. But the overall counting rates of this detector are restricted to the same limits as in the case of the corresponding one dimensional detector, since the same arguments can be applied. In many cases this means a significant drawback. The opposite solution is to omit the delay-lines by reading the signal of every single strip independently. This leads to a drastic increase of the maximum counting rates, but also of the amount of electronics needed. It is possible to find designs which are situated between the two extremes [25].

2.3.2 Vidicon Systems

Silicon Intensified Target (SIT) Vidicons can also be used as two-dimensional detectors [26]. The X-ray photons first have to be converted into visible light. At the HASYLAB this is done by a 30 μm thick layer of Gd_2O_2S, which is deposited on a fibre optic plate. This plate is directly coupled to the photocathode of the vidicon to minimize the intensity losses. At the photocathode, the converted photons produce free electrons, which are accelerated in a focussing electrostatic field and produce an amplified image of the free electron distribution at the photocathode on the silicon target. This image is stored and accumulated, until it is read out by means of an electron beam, analogous to a TV tube. The Vidicon used in Hamburg has an active area of 80 mm in diameter, the picture consists of 512×512 pixels, the spatial resolution is about 0.3 mm, the time between two successive readouts is 40 msec. The maximum photon rate is mainly limited by the converter screen. While a Gd_2O_2S screen is appropriate for observing weak scattering, a ZnS scintillator can work with higher photon rates. 2.1×10^5 photons $\cdot$ sec^{-1} have been registered per TV-scan line without any saturation effects.

2.3.3 Photodiode Arrays

Photodiode arrays are one-dimensional detectors which consist of a linear array of photodiodes, produced in MOS technology on a single silicon chip [27, 28]. Arrays with up to 1024 diodes of 25.4 μm length and 2.5 mm width are presently available, leading to an active length of up to 25 mm.

The reversed-biased p-n junction of each diode acts as a capacitor and collects the charges which are released from the valence band by the photons, which reach the charge depleted area of the diode. The storage and integration process is limited by the leakage current, which can be reduced by cooling the device below −60 °C to obtain integration times of several minutes. Finally the diodes are successively discharged by connecting them, by turns, to a charge sensitive amplifier. This is realized by electronic switches, which are activated by a shift register. The electronics are integrated on the silicon chip, together with the photodiodes.

The diodes are sensitive to a wavelength range from 0.1 nm to 1.2 nm, i.e. they can operate directly with 8 keV photons. However, shielding must be provided to prevent the X-rays from damaging the integrated electronic circuitry. Visible light has to be excluded from the detector, this can be done with a beryllium window.

A photodiode array has a large dynamic range. Values of more than 10^4 can be achieved. The shortest integration times are limited by the speed of the readout electronics. 2.7 msec can be reached for a 1024 pixel array and a 10 bit analog to digital conversion. Another remarkable feature are the high photon rates that can be measured. Each pixel can accumulate about 6×10^4 photons. If this pixel is read out 250 times per second, a maximum count rate of 1.5×10^7 photons per pixel and second results.

The high spatial resolution of the detector of about 26 μm can normally not be utilized at present. With the focussing instruments, the size of the primary beam focus is usually much larger than the width of a single pixel.

2.3.4 Charge Coupled Devices

Since in a photodiode array every single diode has to be addressed by the readout system, only one-dimensional arrays are presently available. In a Charge Coupled Device (CCD) [29] the adjacent MOS-capacitors are put sufficiently close to each other, so that their charge depleted regions overlap. By choosing the potentials of the neighbouring capacitors in an appropriate manner, one can transfer the charges collected at one capacitor to the adjacent one. In this way the charge packets collected along a line of capacitors, can be shifted sequentially into a charge sensitive amplifier connected to the end of the line. This technique reduces the readout electronics markedly and makes it possible to arrange several parallel lines of capacitors, to obtain a two dimensional detector.

2.4 Ancillary Equipment

A great number of devices have been constructed in order to perform certain treatments on the sample, while simultaneously the X-ray scattering is registered. Examples are furnaces, stretching machines, cells for chemical reactions, cryostats etc.

Figure 6 shows a furnace which is widely used at HASYLAB for dynamical measurements during thermal treatment of film shaped polymer samples. It consists of a copperblock with a borehole in its center, through which the X-ray beam can pass through the sample. The metal block can be heated to a given temperature, before the sample, mounted in a small metal holder, is pneumatically injected into it. Heating rates up to 150 °C/min can be achieved, with 500 °C being the maximum temperature. Cooling rates of the same order can also be obtained by leading a cold gas flow through channels, which are provided in the copper block for this purpose. By means of computer control, any temperature program within the given limits can be realized.

3 Results of X-Ray Scattering in Polymers

3.1 Kinetics of Isothermal Crystallization

3.1.1 Polyethylene

For the observation of structural changes during crystallization, recrystallization, and melting of polymers SAXS and WAXS are powerful tools. However, with

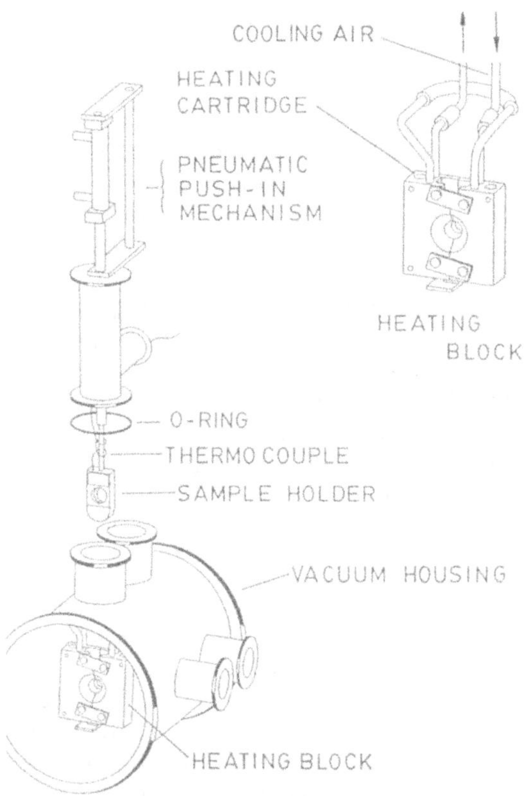

Fig. 6. Furnace used for thermal treatment of samples during SAXS/WAXS measurements. The device is especially designed for rapid temperature changes (150 °C/min.)

conventional X-ray sources in situ investigations of these changes are often impossible, due to the long data acquisition times necessary. A classical way to circumvent this problem is to freeze in the considered process at different times and to examine the intermediate states by static techniques. In the case of crystallization processes, this is often possible, because they stop immediately when the temperature decreases below the glas transition temperature of the material. However, the danger of producing artificial structures by this kind of experiment is obviously extremely high, not least because the freezing cannot happen instantaneously.

S.R. often for the first time opens up the possibility of performing time resolved measurements. It shall be emphasized, however, that the synchrotron light could produce artifacts, too. The question of molecular degradation due to the intensive primary X-ray beam must be carefully noticed, especially when the measurements are carried out at elevated temperatures above glass transition.

SAXS measurements of the lamellae thickness of solution crystallized polyethylene (PE) as a function of supercooling $\Delta T = T_m - T_c$, where T_m and T_c are the melting and crystallization temperatures respectively, are in good agreement with the theory. It predicts, that the thickness of the lamellae, l_c, is determined by the size of the critical nucleus, which is given by [30]

$$l_c = \frac{2\sigma_e T_m}{\Delta H (T_m - T_c)} . \tag{5}$$

T_m is the melting temperature of the infinitely extended crystal, ΔH is the change in enthalpy when melting a single monomer layer of the crystal, and σ_e is the surface free energy of the nucleus. In the case of crystallization from the melt, however, the thickness was always found to be too large. Barham et al. [31] and Martinez-Salazar et al. [32], for the first time, performed time-resolved X-ray scattering studies during melt crystallization of PE. They found that the long period at the beginning of the crystallization process is much smaller than it had been found in quenched samples. Shortly after the formation of these primary lamellae, a very rapid thickening takes place, leading to an increase of the thickness by a factor of about two. After this sudden increase the lamellae grow further logarithmically with time, which was already known before and can be explained by a longitudinal diffusion of the chains within the crystals. Therefore, by means of S.R. it was possible for the first time to determine the true primary crystal thickness as a function of supercooling ΔT. The experiments establish that it is this thickness, which is defined by the size of the critical nucleus and its dependence of ΔT. Now the experimental results and the theory are in good agreement for melt crystallized PE, too.

3.1.2 Polyethylene Terephthalate

Polyethylene terepthalate (PET) differs from PE mainly by a benzene ring, which is incorporated into every monomer unit of the chain molecule. Compared to PE, the chain becomes bulkier, which influences the crystallization behaviour of this polymer. PET can be obtained perfectly amorphous when quenched from the melt. Thus it can be crystallized from the amorphous solid state, in contrast to PE, which is always obtained with a high degree of crystallinity. The degree of crystallinity that is reached in PET depends on many material parameters and crystallization conditions, but more than 60% is unusual.

Elsner et al. [33] followed the SAXS during isothermal crystallization of initially amorphous PET. Figure 7 shows the temporal development of the scattering at 125 °C. First an increase of the diffuse scattering is observed, as a consequence of the formation of crystalline domains, which do not have a significant spatial order relative to each other. After about 1.8 min. a shoulder emerges indicating that the crystal lamellae start to form regular stacks. At subsequent times the position of the maximum shifts gradually towards larger angles. The long period thus decreases which is in contrast to the behaviour observed with PE.

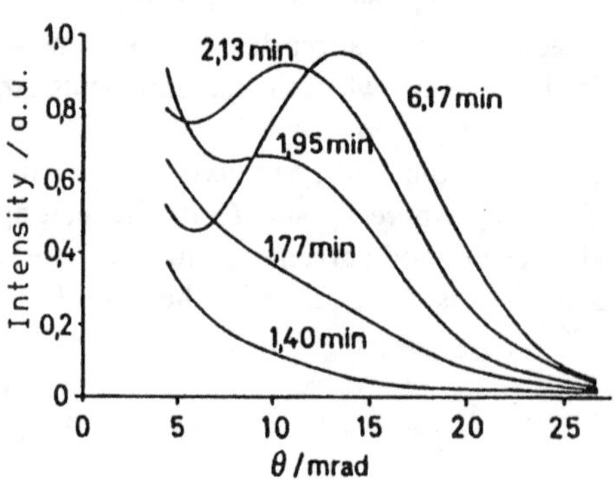

Fig. 7. Small-angle X-ray scattering curves of polyethylene terephthalate after different times of isothermal crystallization from the amorphous state at 125 °C

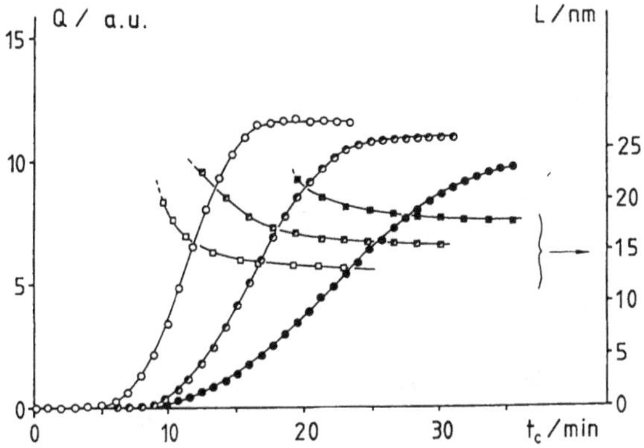

Fig. 8. Scattering power Q and long period L as a function of time t_c during crystallization of PET from the melt at 235 °C; (Parameter is the molecular weight, ○ □: M = 21000, ◐ ◩: M = 33000, ● ■: M = 48000)

The isothermal crystallization of PET was systematically investigated by Gehrke et al. [34, 35, 36] by SAXS and WAXS using S.R. Figure 8 shows, as a result, the scattering power Q and the long period L obtained from SAXS as a function of time during a crystallization at $T_c = 235$ °C, coming from the molten state. The parameter is the molecular weight M of the sample. The long period appears after 25% of the material has already been crystallized into spherulites. During the further growth of the spherulites the already mentioned decrease of the long period takes place. From Q the degree of cyrstallinity w_{cs} at the end of the crystallization can be determined according to Eq. (3). Assuming a two phase system, the thickness of the crystalline lamellae, l_c, and of the amorphous layers, l_a, can be estimated to be

$$l_c = w_{cs} L \,, \qquad l_a = (1 - w_{cs}) \, L \,. \tag{6}$$

This relations are only correct, if all amorphous material is solely located between the crystal lamellae within the stacks. In Fig. 9 l_c and l_a at the end of crystallization are plotted as a function of the crystallization temperature T_c. Up to $T_c = 235$ °C l_a is almost temperature independent, while l_c increases steadily. This increase is in a good agreement with the theoretical predictions. Like the initial crystal thickness in PE also the thickness of the lamellae in PET is defined by the size of the critical nucleus. But in PET these initial lamellae do not grow further. Instead of this, the decrease of the

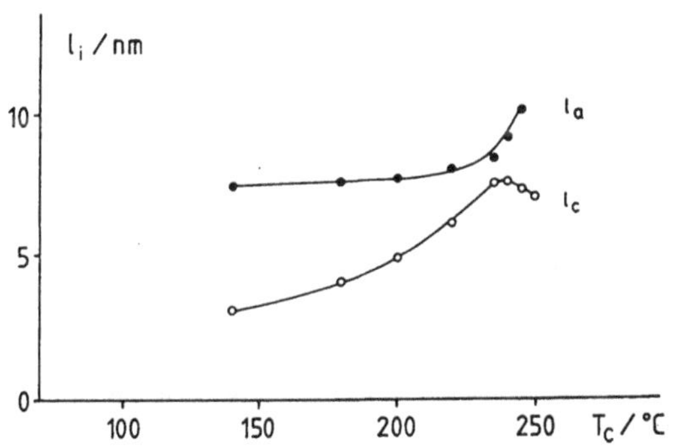

Fig. 9. Thickness of the amorphous regions (l_a) and crystalline regions (l_c) at the end of a crystallization of PET from the melt as a function of the crystallization temperature T_c

long period indicates, that the mean distance of the lamellae becomes smaller during crystallization.

The thickness of the amorphous regions is probably defined by the minimum distance, two independently growing nuclei can have to each other. This value is obviously not influenced by the temperature. Above 235 °C, l_a seems to increase at expense of the crystalline regions. The sample is in a partially molten state, which leads to the existence of amorphous islands. It could be shown that these islands are filled with smaller lamellae, when the temperature is decreased so far that they can become thermodynamically stable.

Elsner et al. [37] observed the change of SAXS during isothermal crystallization of PET from the glassy state. The samples were previously oriented by drawing the amorphous material. In order to register the anisotropic scattering patterns, a vidicon system was used. Figure 10a shows the long period L as a function of the crystallization time t_c for different temperatures T_c. As with the unoriented samples, an initial decrease of L is observed. The corresponding azimuthal half widths $\Delta\varphi$ of the SAXS reflection is given in Fig. 10b. This value directly characterizes the width of the orientation distribution of the normals onto the surfaces of the crystal

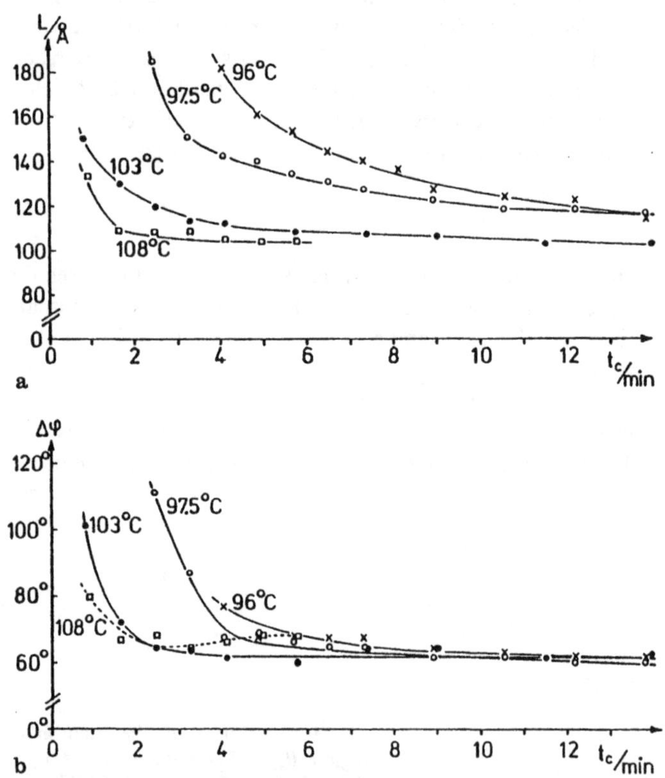

Fig. 10. Long period L and azimuthal half width $\Delta\varphi$ of the SAXS maximum as a function of time t_c during a crystallization of oriented PET (initial birefringence $\Delta n_0 = 0.19$). Parameter is the crystallization temperature

lamellae. One can see, that the degree of orientation of the surfaces improves during crystallization. Other measurements show, that the azimuthal width of the corresponding WAXS reflections are constant from the beginning. This means a constant orientation distribution of the molecules during the crystallization process. From this observations the conclusion was drawn, that at the beginning of the crystallization process the lamellae are corrugated as sketched in the left part of Fig. 11. During the crystallization, the lamellae flatten by parallel shifting of the chains (see right side of Fig. 11). This implies that the orientation of the chains remains unchanged, while the orientation of the lamellae surfaces is improved and the mean distance of the lamellae decreases.

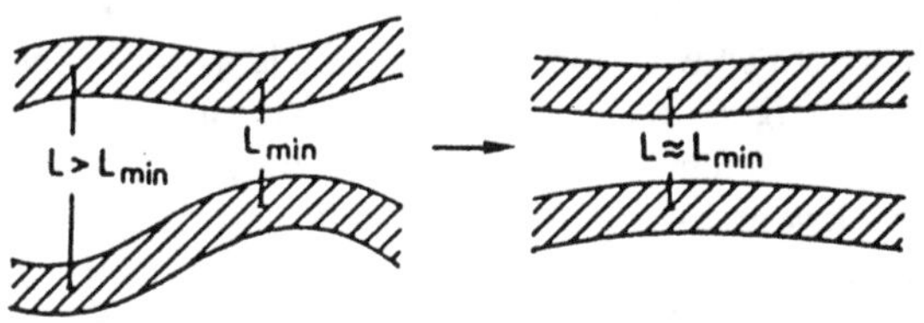

Fig. 11. Proposed model of flattening of the initially corrugated crystal lamellae during crystallization

3.1.3 Copolyester of Poly-β-hydroxybutyrate and Polyethylene Terephthalate

By means of SAXS and WAXS the crystallization of copolyesters consisting of poly-β-hydroxybutyrate (PHB) and polyethylene terephthalate (PET) was investigated. Samples with different contents of PET were synthesized and crystallization was followed at various temperatures [38]. The aim was to study, how an increasing fraction of the inflexible PHB-groups influences the crystallization kinetics of the copolyester. The measured half times of the crystallization, i.e. the time interval, during which the crystallinity reaches the half of its value at the end of the main crystallization, are shown in Fig. 12 as a function of temperature. The parameter gives the weight fraction of PET. It is known, that the glass transition temperature T_g, i.e. the temperature above which the chain segments are able to exchange their locations, is nearly independent of the PHB content. Thus the

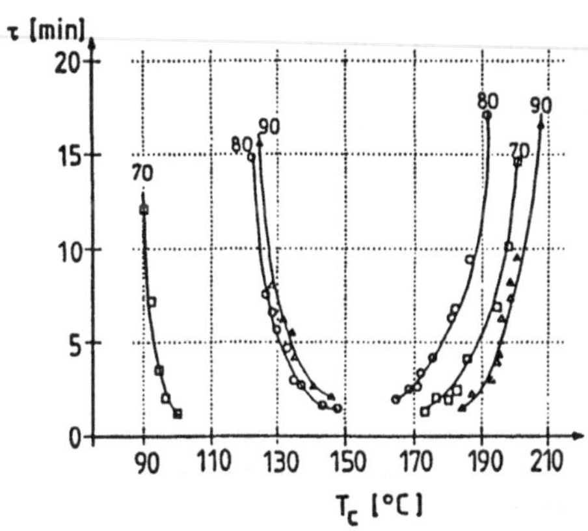

Fig. 12. Half times τ of the isothermal crystallization of PET:PHB copolyesters as a function of crystallization temperature T_c. Parameter is the weight fraction of PET

observed differences in the half times in the low temperature range cannot be explained by different T_g values leading to an onset of crystallization at different temperatures. Thus the observation must be attributed to the inflexibility of the PHB groups, resulting in an increase of the crystallization velocity with increasing PHB content. In the high temperature range two competitive effects might influence the half times: the change of inflexibility and the change of the melting temperature.

3.1.4 Polyethylene naphthalene-2,6-dicarboxylate

Polyethylene naphthalene-2,6-dicarboxylate (PEN) has a glas transition temperature of T_g = 120 °C and the crystals melt at T_m = 280 °C. In order to obtain completely amorphous samples, one can melt the material for several minutes at 330 °C and then quench it below T_g. Such amorphous samples were crystallized and the WAXS followed almost simultaneously. Figure 13 shows some of the results [39], on the left side obtained for T_c = 167 °C, on the right for T_c = 245 °C. Obviously different crystal modifications are formed at the two temperatures. The half time of crystallization was found to be 6 min for the low and 4 min for the high T_c. In none of the experiments a transformation from one modification to the other was ever observed. The low temperature modification seems to be formed more easily and after melting below 300 °C this modification even crystallizes in the temperature region, where usually the high temperature modification should appear. Only if the low temperature modification has been molten above 330 °C, then after cooling to the appropriate temperatures, is the high temperature modification formed. From this observations one can conclude, that the nuclei of the low temperature modification are more easily formed and that they are not perfectly destroyed when melting below 330 °C. On the

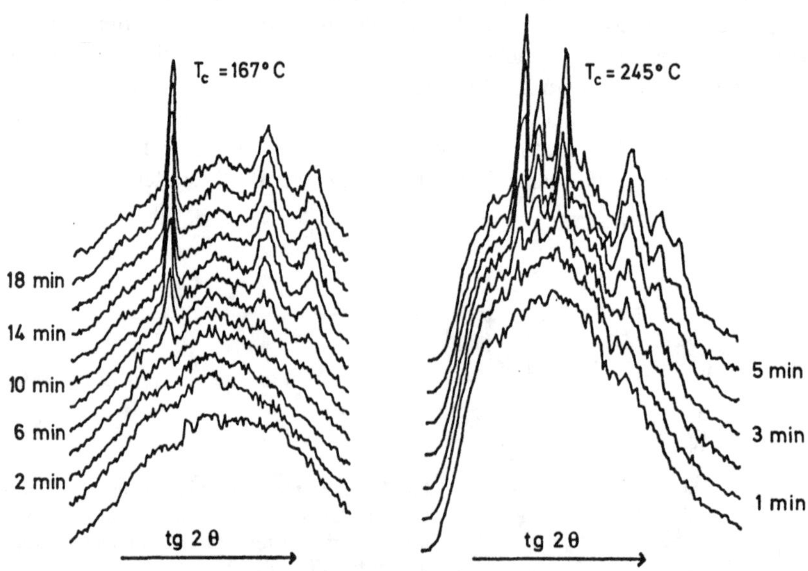

Fig. 13. Change of wide-angle X-ray scattering during crystallization of initially amorphous PEN at two different temperatures T_c coming from the glassy state

other hand the high temperature modification is more difficult to nucleate, but obviously thermodynamically more stable.

3.1.5 Isothermal Crystallization and Spinodal Decomposition

With regard to the crystallization of polymer material from the amorphous state it is usually assumed, that the growth of crystals starts from a nucleation process [30]. If due to thermal fluctuations a nucleus with dimensions beyond a critical size is formed, it will be thermodynamically stable and will become the center of a growing crystal. Especially for the crystallization of highly oriented systems an alternative model has been proposed [40], which is similar to the model of spinodal decomposition of binary systems [63]. It is based on the idea, that an amorphous phase is characterized by a high concentration of chain segments with kink-like defects. Accordingly, the crystals are regions, which exhibit low defect concentrations. The decomposition model, based on free energy arguments, proposes that at least in oriented systems the kink-defects could diffuse along the chains, leading to regions of increased defect concentration, while around these regions the defect concentration decreases. This process should gradually lead to sinoidal density fluctuations, which are initially described by a superposition of a broad spectrum of different wavelengths. During the progress of this local defect concentration a certain wavelength should be continuously amplified, while the others gradually vanish. A characteristic feature of this process is the continuous increase of the density difference between the crystalline and amorphous regions, $\varrho_c - \varrho_a$, in contrast to the crystal growth after nucleation, where this difference will be constant.

In order to examine, whether a change in $\varrho_c - \varrho_a$ takes place during the main crystallization of unoriented amorphous PET, Prieske et al. performed simultaneous measurements of time resolved SAXS and WAXS [41]. The WAXS was registered on a photographic film, which was changed automatically at time intervals of 60 sec. In the center of the films, holes of 2 cm in diameter were provided in order to permit

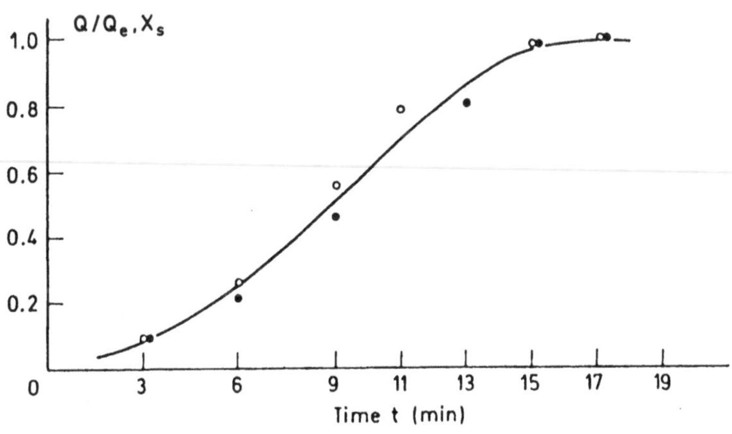

Fig. 14. Relative scattering power Q/Q_e (○ ○) and fraction of spherulitic crystallized materials x_s (● ●) as a function of crystallization time t during isothermal crystallization of PET from the glassy state at 117 °C

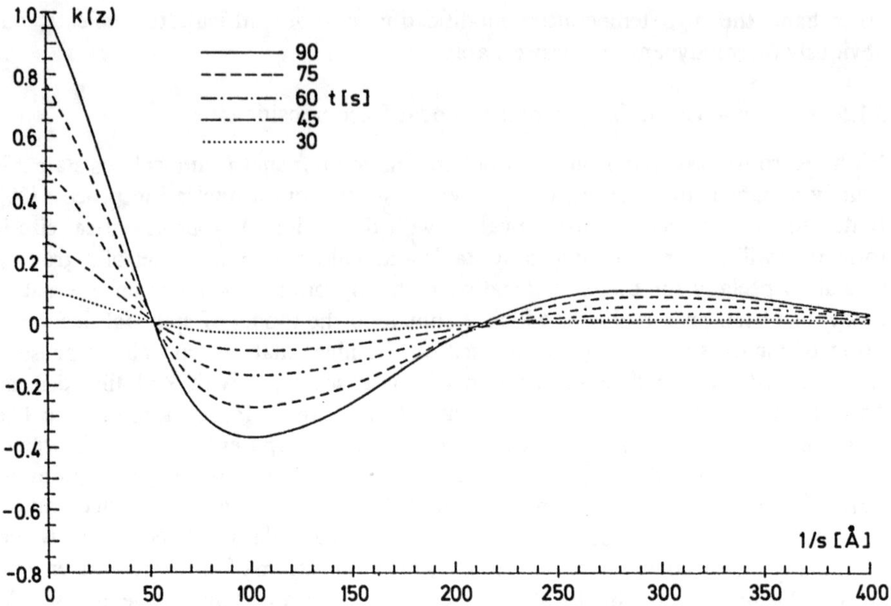

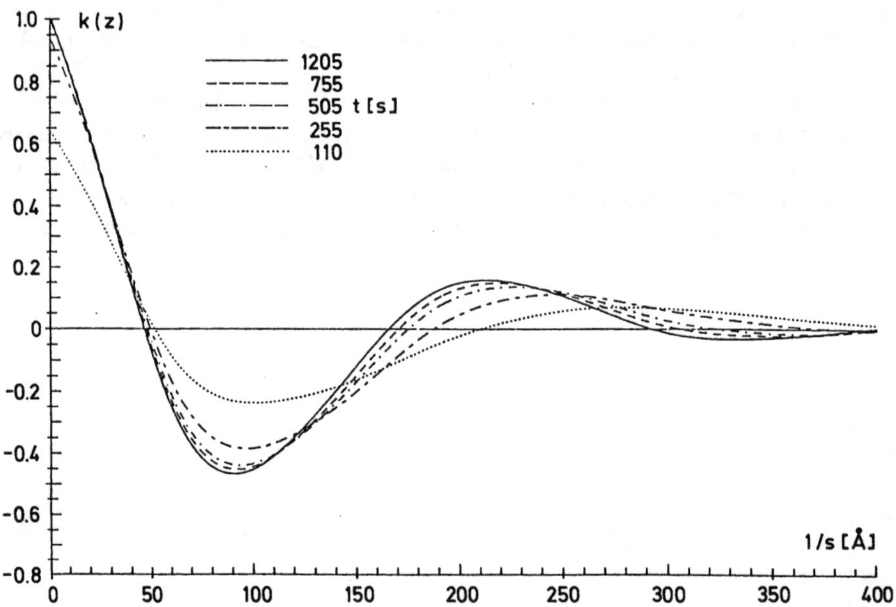

Fig. 15. Scattering density correlation functions k(z) at different times t during crystallization at $T_c = -4.5$ °C of an oriented crosslinked *cis*-polybutadiene sample which was previously molten in the drawn state ($\lambda = 5.6$)

the SAXS to pass through the film. The SAXS was then registered on a one-dimensional gas filled detector placed 3 m behind the sample position. From the WAXS the volume fraction of spherulitic crystallized material, x_s, can be directly determined as a function of time. In Fig. 14 the results obtained for a crystallization at 117 °C are plotted, together with the scattering power Q as determined from the simultaneously recorded SAXS patterns. Both curves exhibit an identical course. With Eq. (3) one can conclude from this result, that $\varrho_c - \varrho_a$ does not change during the crystallization process. The observed increase of Q is purely caused by a corresponding increase of x_s. The crystallization thus takes place by nucleation and subsequent crystal growth, and not by a spinodal decomposition type mechanism.

Different results were found by Fischer et al. in oriented systems [42]. Peroxyde crosslinked *cis*-polybutadiene was initially molten in a drawn state. Then it was rapidly cooled to the crystallization temperature and the SAXS or WAXS was followed during isothermal crystallization. The time resolution was about 5 sec. The corrected scattering curves were fourier transformed, leading to the scattering density correlation functions [10] which can be analyzed further. As an example of the different experiments, the correlation functions at different times during a crystallization at $T_c = -4.5$ °C are given in Fig. 15. The draw ratio was $\lambda = 5.6$, the parameter is the crystallization time. One can see, that for short times the crystallization process can be described by a continuous increase of the density differences, with the distance of the density maxima remaining constant. The time dependence of the maximum density value is consistent with the theory of a spinodal defect decomposition. For long times (i.e. $t_c > 100$ s) deviations from this picture appear. They can be explained by the additional formation of new crystal lamellae in the amorphous regions between the already existing ones. This leads to a decrease of the distances of the correlation function maxima.

3.2 Kinetics of Recrystallization

Crystal lamellae, which have been formed at a given temperature T_1 become thermodynamically unstable at sufficiently elevated temperatures, $T_2 > T_1$. The higher temperatures correspond to the formation of lamellae of increased thickness. Thus usually a transformation occurs, when a sample is annelated above the crystallization temperature [15]. This process is called recrystallization and a general question comes with it: does the increase of the lamellae thickness take place by a continuous longitudinal diffusion of the chains within the crystals or do the old lamellae first melt before new ones of increased thickness are formed?

3.2.1 Polyethylene

Grubb et al. [43] studied this question on single crystal mats of polyethylene. The samples, previously crystallized at 80 °C, were annealed in a flow of hot helium at different temperatures above 120 °C. WAXS was observed during recrystallization by means of a vidicon detector, the time resolution being 0.3 sec. A strong decrease of the intensity of the (110) crystal reflection and therefore of the crystallinity is observed during the first few seconds of the experiment. Simultaneously a rapid in-

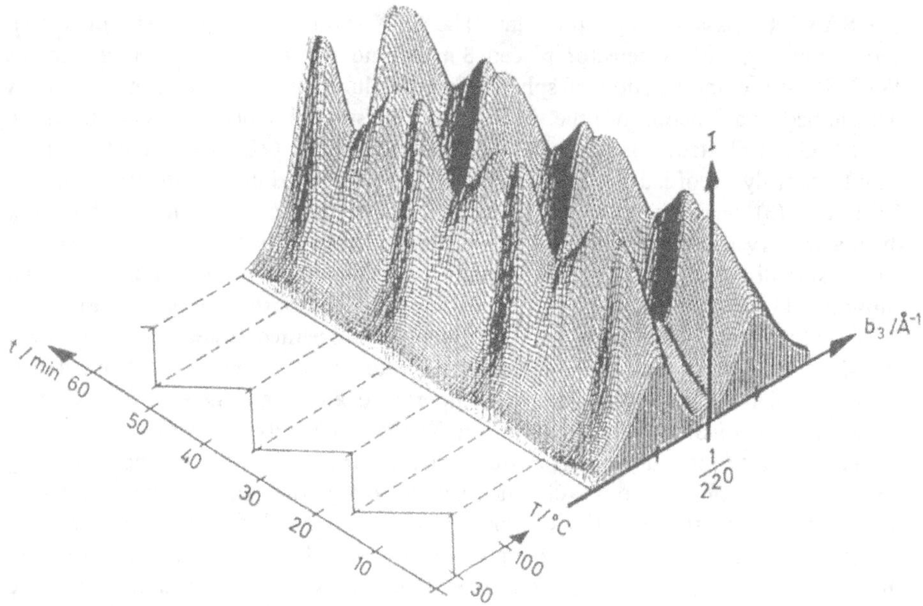

Fig. 16. Change of the small-angle X-ray scattering pattern of oriented low density polyethylene during four temperature cycles. The scattering was recorded by a linear position sensitive detector oriented parallel to the meridian of the scattering pattern

crease of the long period is found. The results indicate, that in the case of annealing of PE, the recrystallization takes place by melting and subsequent formation of new crystals.

Fronk et al. [44] found reversible changes of the long period of oriented PE, when the temperature was elevated into the melting range and subsequently was decreased again. Figure 16 testifies this experiment, in which the temperature was alternately changed between 30 °C and 100 °C. The SAXS pattern has been measured symmetrically to $s = 0$ by means of a linear detector. The long period changes reversibly between 21.7 nm at the low, and 25.2 nm at the high temperature. This can be explained by the coexistence of lamellae of various thickness. At the high temperature the thin lamellae become unstable and melt. Since the small lamellae are situated between thick ones, which remain stable, this melting results in an increase of the mean lamella distance. During cooling, the thin lamellae can recrystallize again.

3.2.2 Polyethylene Terephthalate

Recrystallization of PET was studied by Gehrke et al. [34, 35, 36]. Samples, previously crystallized at 120 °C, were rapidly heated up to higher temperatures, with rates of 100 °C/min. During annealing the SAXS was observed by means of a linear detector, the time resolution of the measurements being 20 s. Figure 17 shows the scattering curves after different annealing times t_a at 250 °C. First the reflection maximum shifts rapidly towards lower angles, before the sample has attained its final temperature. Soon after 250 °C is reached, the reflection maximum vanishes

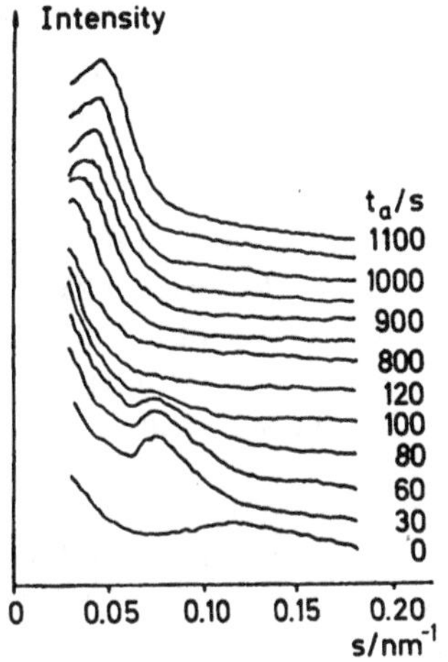

Fig. 17. Small-angle X-ray scattering patterns of PET previously crystallized at 120 °C at different times t_a during annealing at 250 °C

completely (t_a = 120 s), then it appears again at even smaller angles, before it shifts gradually towards higher ones. From this observation it is evident, that at such high temperatures old crystals melt before the formation of new ones. Figure 18 shows adequate measurements at T_a = 235 °C. Here the maximum shifts gradually to lower angles and there is no direct evidence for any melting. This could indicate a continuous thickening of the crystals, but the observations could also be explained by melting and recrystallization, if it is assumed, that the two concurrent processes are superimposed, resulting in a gradual shifting of the mean long period. This assumption seems to be plausible, since the crystallization is known to be much faster at 235 °C than it is at 250 °C.

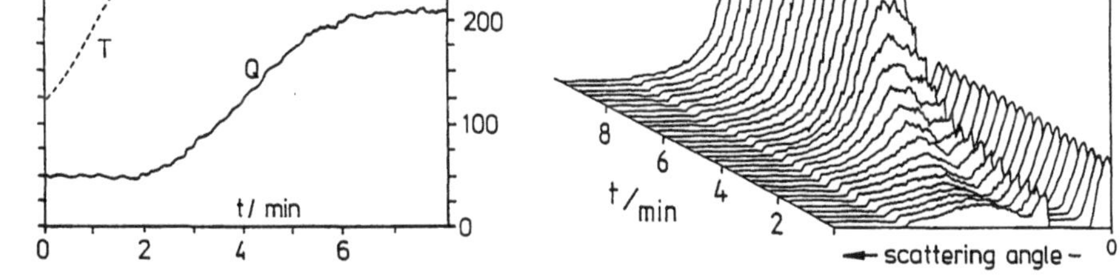

Fig. 18. Small-angle X-ray scattering patterns of PET previously crystallized at 120 °C during heating to 230 °C. On the *left side* the corresponding time course of the long period L, scattering power Q, and temperature T are given

In order to further illuminate this problem, the kinetics of this recrystallization processes were studied in more detail. The SAXS was followed during annealing and from the scattering power Q the degree of crystallization, w_{cs}, was determined according to Eq. (3). Comparative WAXS measurements were necessary to calibrate the crystallinity measurements, since with Q only relative crystallinity values are obtained. In Fig. 19 w_{cs} is plotted as a function of the annealing time for different molecular weights M and two different annealing temperatures T_a. The curves at 250 °C show directly melting and recrystallization as two consecutive processes, whereas at 235 °C w_{cs} increases continuously. Remarkable is the dependence of the recrystallization velocity on the molecular weight. This suggests that the whole molecule is involved in the recrystallization process and it thus seems to indicate, that the recrystallizing molecules are initially in a nearly completely molten state. This assumption is supported by an analysis of the thickness of the crystalline and amorphous regions at the end of the recrystallization, as they are determined from SAXS according to Eqs. (3) and (6). Figure 20 shows l_c and l_a as a function of the annealing temperature T_a. The dependence on T_a is quite similar to the corresponding dependence of l_c and l_a on T_c after crystallization from the melt (compare with Fig. 9). The constant amorphous thickness, l_a, which is characteristic for the melt crystallization, is difficult to explain in the case of recrystallization, if a continuous crystal thickening is supposed, since this thickening should be at the expense of the amorphous regions.

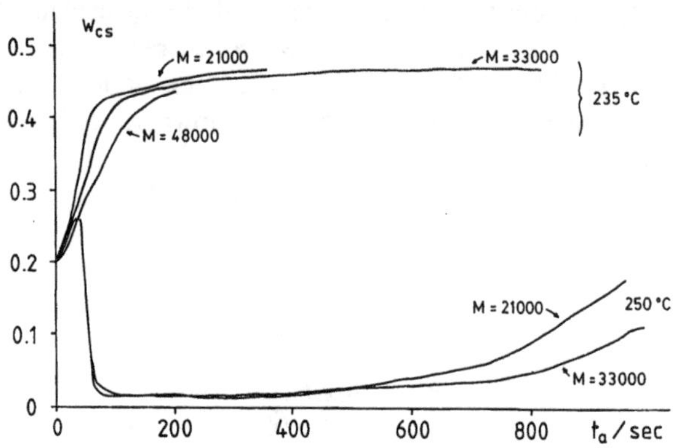

Fig. 19. Degree of crystallinity w_{cs} of PET as a function of annealing time t_a of a sample previously crystallized at 120 °C. Parameters are the molecular weight M and the annealing temperature T_a

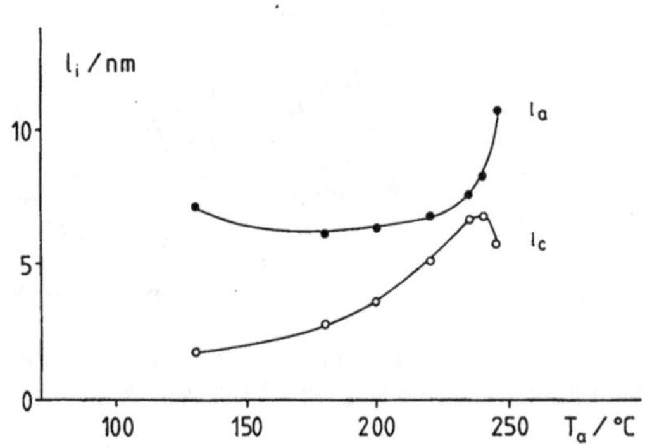

Fig. 20. Thickness of the crystal lamellae l_c and of the amorphous regions l_a in PET at the end of annealing as a function of annealing temperature T_a. The samples were previously crystallized at 120 °C

Significantly different results are obtained, if the material is heated up slowly. Figure 21 shows l_c and l_a, determined from SAXS as a function of temperature during heating the sample with constant rates from the initially amorphous state. The result of three different heating rates are plotted. The crystal thickness, l_c, exhibits nearly the same course for all rates. The value is obviously defined by the equilibrium thickness corresponding to each temperature. For fast heating rates, this equilibrium states are reached in the previously discussed manner by melting and recrystallization, the amorphous regions remaining constant. At low heating rates the crystal growth reduces the thickness of the amorphous regions. Obviously, during slow heating enough time is available for the crystal lamellae to reach the required thickness by continuous growing. It is not clear, whether this happens due to chain diffusion or due to rearrangements of the chains at the lamellae surfaces, the latter being facilitated by transesterification processes. If, as a consequence of a high heating rate, not enough time remains for this continuous process, then melting with consecutive recrystallization becomes dominant.

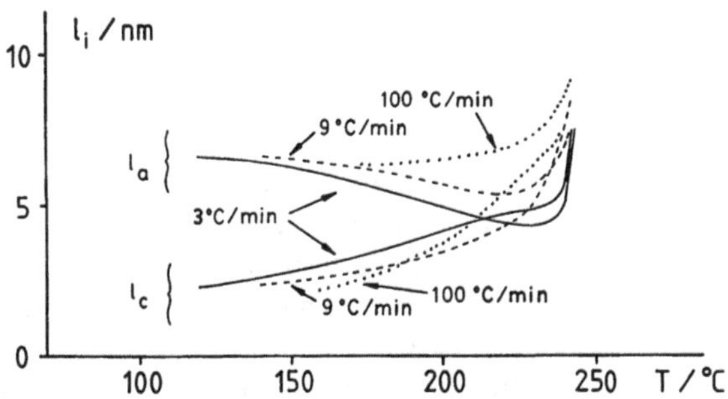

Fig. 21. Thickness of the crystal lamellae l_c and of the amorphous regions l_a in PET as a function of temperature T during heating of initially amorphous material with different constant heating rates

3.3 Stress Induced Crystallization

Polyisobutylene forms crystalline domains when the material is stretched. Studies of the kinetics of this stress induced crystallization were the motive for the first

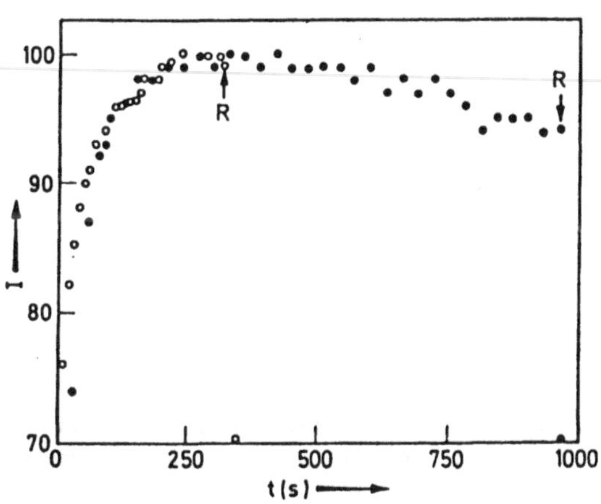

Fig. 22. Intensities of the (020) (● ●) and (113) (○ ○) reflections as a function of time after rapid stretching of poly isobutylene fibres. Upon releasing the stress (R) the intensities immediately drop back to the initial values

application of time resolved WAXS measurements using S.R. [45]. The samples were elongated at room temperature to draw ratios between 300% and 800% within a fraction of a second. In Fig. 22 the intensities of the (020) and (113) crystal reflections are plotted as a function of time after elongation. This reflections emerge immediately after the stretching has taken place and a half time of crystallization of 40 sec. could be estimated. The results are explained by an athermal formation of fibrils. When the stress is released, the reflections disappear immediately. Supplementary studies [46] proved, that stress induced crystallization also takes place at drawing velocities down to 0.1 mm/min.

Caffrey et al. [47] performed similar studies on natural rubber. Again stress induced crystallization is observed and the WAXS reflections disappear, when the material relaxes. More detailed studies on the reversibility of this process were performed by Holl et al. [48].

3.4 Transformation of Crystal Modifications

3.4.1 Poly-β-hydroxybutyrate

Poly-β-hydroxybutyrate (PHB) crystallizes in a low temperature modification (LTM), which is transformed into a high temperature modification (HTM) when heated above 320 °C. Both modifications exhibit distinct WAXS patterns, thus a quantitative determination of the two fractions is possible. However, measurements at the HTM are only feasible with S.R., since the molecules quickly decompose at the high temperatures. Figure 23 shows the fraction of the HTM during heating the PHB sample from 280 °C to 340 °C at a rate of 2 °C/min and subsequent cooling to 280 °C at the same rate [49]. A reversible transformation is observed. But the transition from the LTM to the HTM occurs at a temperature, which lies about 20 °C higher than the transition in the opposite direction. This indicates, that each modification is formed by its own nucleation process.

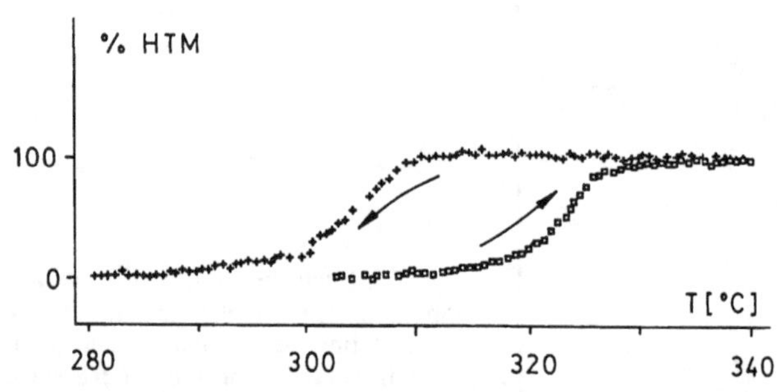

Fig. 23. Fraction of the high temperature modification (HTM) in PHB during heating and subsequent cooling with constant rate across the temperature region where a modification transformation occurs

3.4.2 Polyacetylene

Riekel [50] investigated the irreversible transformation of *cis*-polyacetylene into th *trans*-modification. The sample was heated from room temperature to 350 °C wit a rate of 11 °C/min and the WAXS was recorded continuously. The result is shov in Fig. 24. At 150 °C the (201) crystal reflection vanishes, indicating the *cis/trans* transition. Above 290 °C a decomposition takes place, leading to the observed bro peak.

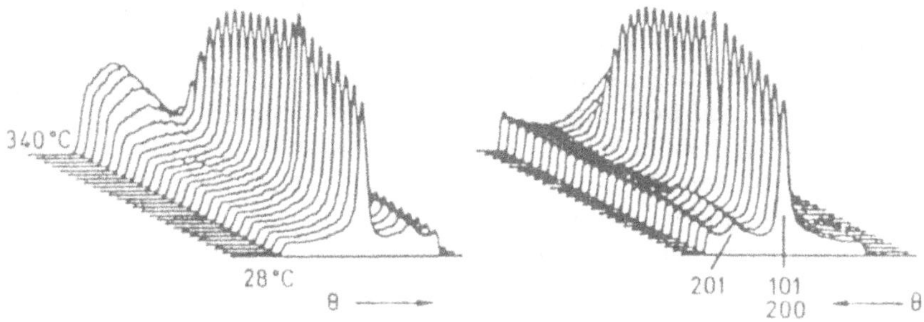

Fig. 24. Change of the wide-angle diffraction pattern of *cis*-polyacetylene during heating from room temperature with 11 °C/min. The disappearance of the (201) reflection at 150 °C indicates the *cis/trans* isomerization, the broad peak above 290 °C indicates thermal degradation

3.5 Simultaneous Measurement of X-Ray Scattering and Heat Transfer

Useful information about melting, crystallization, and other phase transitions can be obtained by measuring the heat transfer between the sample and its surrounding, while the transformation takes place. Differential Scanning Calorimetry (DSC) is thus often applied in addition to time resolved X-ray scattering measurements [51]. Since it is very difficult to conduct two different experiments in exactly the same manner, it is desirable to perform both experiments simultaneously. The idea to expose the sample within a DSC cell to a collimated X-ray beam by means of a small window and to register the scattered radiation at the other end, was first realized by Russell et al. [52]. A commercially available DSC cell designed for optical microscopic studies was modified for this purpose. The calorimetric signal during increasing the temperature of a PE sample was registered, together with the SAXS measured by means of a photodiode array.

The same authors investigated the melting behaviour of polyurethanes [53], which are multicomponent polymers, exhibiting at least three different phases and a multiple endothermic response during heating. By comparing the SAXS and DSC data obtained during a continuous temperature increase, it was found that at first small crystals melt and mix with the amorphous soft segment phase. At higher temperatures finally the larger hard segment crystals melt, too.

Recently Bark et al. [54] have used a similar DSC cell to measure heat transfer, SAXS, and WAXS simultaneously. In this experimental design the WAXS was registered by a linear gas filled detector in a range of scattering angles from $\theta = 7°$

to $\theta = 18°$. The SAXS in an interval from $\theta = 0.1°$ to $\theta = 2°$ was permitted to pass beside the detector in order to be detected by a vidicon situated at a larger distance from the sample position. This arrangement was utilized to investigate the crystallization of PEN. An initially amorphous sample was heated at a constant rate. Figure 25 shows the heat transfer dH/dt, together with the scattering power Q determined from SAXS, and the intensity I_{010} of the (010) crystal reflection obtained from WAXS, as a function of temperature. The sample crystallizes in the range from 185 °C to 210 °C, which is indicated by the exothermal DSC signal and by the increase of Q and I_{010}. Above 195 °C, Q increases only due to the different thermal expansion coefficients of the crystalline and amorphous regions, leading to an ascent of $(\varrho_c - \varrho_a)^2$ and therefore of Q, according to Eq. (3). This is established by a detailed analysis of the data. On the other hand, the heat transfer rate and the WAXS intensity I_{010} indicate a further increase of the crystallinity. This observation can be explained assuming the main crystallization to be finished, when 195 °C is reached. At that instant, the sample is completely filled with spherulites, the further crystallinity increase takes place within the spherulites and according to Eq. (3), Q does not rise proportionally to x_s, as during the main crystallization, but proportionally to $w_{cs}(1 - w_{cs})$. Since w_{cs} is close to 0.5, a further change of w_{cs} does not strongly influence Q. Hence the simultaneous measurement of the three quantities allows the direct distinction between the main and the residual crystallization processes. The maximum heat transfer coincides with the end of the main crystallization. At 255 °C the material starts to melt, as can be seen from the endothermic DSC peak and from the decrease of Q and I_{010}. At 280 °C the sample is completely molten.

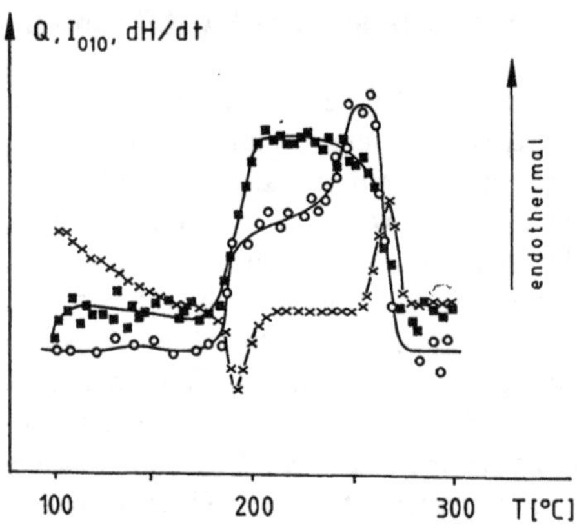

Fig. 25. Heat transfer dH/dt, scattering power Q, and intensity of the (010) crystal reflection I_{010} as a function of temperature T during heating of initially amorphous PEN with constant rate

3.6 Determination of Molecular Orientation by X-Ray Pole Figures

3.6.1 Wide Angle X-Ray Scattering Pole Figures

Mechanical treatment of polymer materials, like stretching, rolling, injection moulding etc. leads to a non-uniform orientation distribution of the crystalline regions. WAXS texture analysis [16] is a well known method of determining the orientation distribution of a certain crystallographic plane.

Let us consider the directions of an incident and an outgoing X-ray beam fixed in space, so, that the Bragg condition Eq. (1) for a certain type of net planes (h, k, l) in a given poly-crystalline material is fulfilled. Then only the fraction of net planes with their normals bisecting the angle between incident and outgoing beam contribute to the reflection intensity measured in the direction of the outgoing beam. Consequently the measured intensity is proportional to the number of net planes oriented in this manner. If the sample is successively rotated to each orientation relative to the fixed beam geometry, an intensity distribution is obtained, which is directly related to the orientation distribution of the normals onto the considered net planes. In order to perform such measurements, the sample is usually mounted on a goniometer as shown schematically in Fig. 26. When measuring the intensity in reflection geometry, the sample is rotated completely about the φ-axis assigned in the figure and the inclination is changed by rotating a small amount about the α-axis, alternately. The measured intensities $I(\alpha, \varphi)$ are corrected according to geometrical effects and are represented in a polar coordinate system (α, φ), with φ being the azimuthal angle. Points of equal intensity are connected in order to obtain a contour map, the pole figure, which directly displays the density of net plane normals with an orientation described by α and φ. Film shaped samples are normally mounted onto the goniometer in such an orientation, that the net plane normals that belong to the center of the pole figure (i.e. to $\alpha = 0°$) coincide with the normals to the film surface. Then the normals that correspond to $\varphi = 90°$ are parallel to the film surface and φ is the angle included by the net plane normal and a preferred direction (transverse direction, TD) of the sample.

Due to the large number of sample orientations that must be successively measured, the data acquisition for a complete pole figure is rather time consuming, typical times are 24 hours with conventional X-ray sources. The high intensity of S.R. can reduce the acquisition time to about 3 hours, this time being mainly determined by the time necessary to change the angular positions of the goniometer.

A further improvement of the efficiency can be gathered by using linear detectors. Instead of the pole figure of a single net plane, the pole figures of a continuous

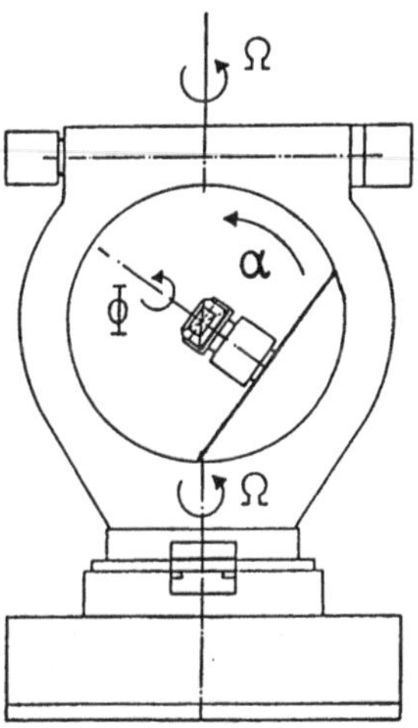

Fig. 26. Schematical representation of a goniometer for the acquisition of X-ray pole figures

interval of scattering angles are measured simultaneously. This especially facilitates the solution of the problem, which arises from the overlapping of adjacent reflections. Due to this overlap, the pole figure of a given net plane often includes contributions from other net planes or from the amorphous halo. The degree of the superposition additionally depends on α and φ. A separation becomes possible if the detailed nature of this superposition at a given α, φ-pair can be examined by means of the intensity profile as a function of the scattering angle θ.

The use of two-dimensional detectors reduces the data acquisition time further, because the rotation about the azimuthal angle φ can be omitted. This leads to measuring times of about 20 min. for a complete pole figure, which additionally includes completely the interesting region of scattering angles.

Röber et al. [55] constructed a pole figure goniometer to be used with S.R. and position sensitive detectors. WAXS pole figures were measured on biaxially drawn

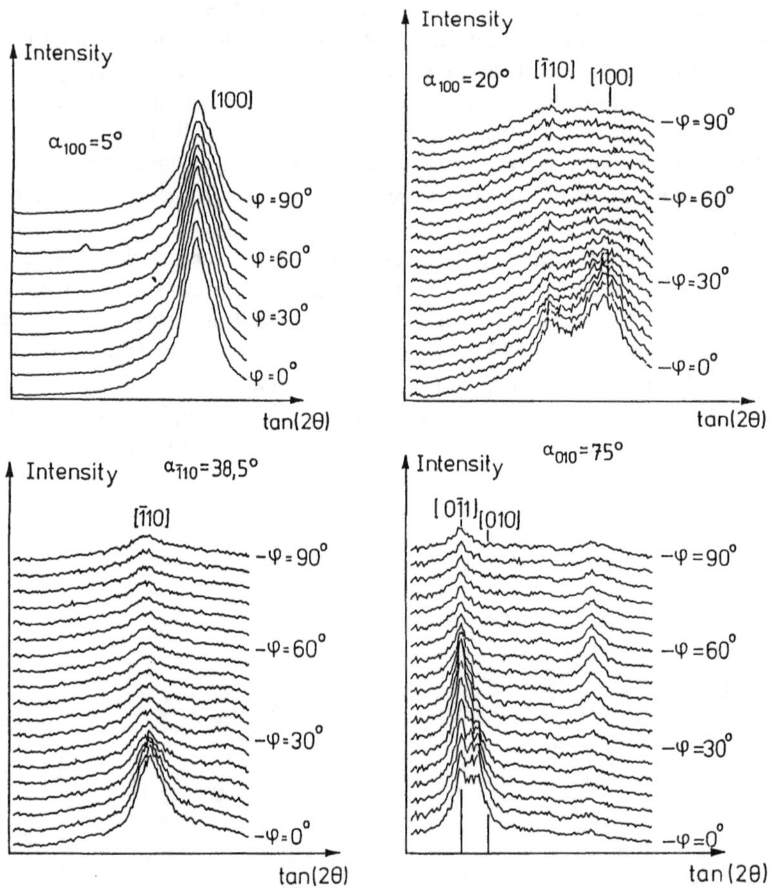

Fig. 27. Wide-angle X-ray scattering profiles as a function of the scattering angle tan (2θ) of a biaxially oriented PET film for various fixed angles α and φ (for the meaning of these angles see Fig. 26)

films of polyethylene terephthalate, i.e. the samples were first drawn in the longitudinal machine direction, MD, and afterwards in the perpendicular transverse direction, TD. Figure 27 shows the scattered intensity as a function of the scattering angle θ for four different values of α and 20 values of φ. The various contributions of the (100), (—110), (0–11), and (010) net planes and of the diffuse amorphous halo can be distinguished.

The pole figure of the (—110) net planes obtained after separation of the adjacent reflections and of the amorphous halo is given in Fig. 28a. The normals of the net planes exhibit a preferred orientation at $\varphi = 0°$ and $\alpha = 37°$. Comparison with Fig. 28b, which is obtained without correction of overlapping reflections demonstrates the necessity of the corrections.

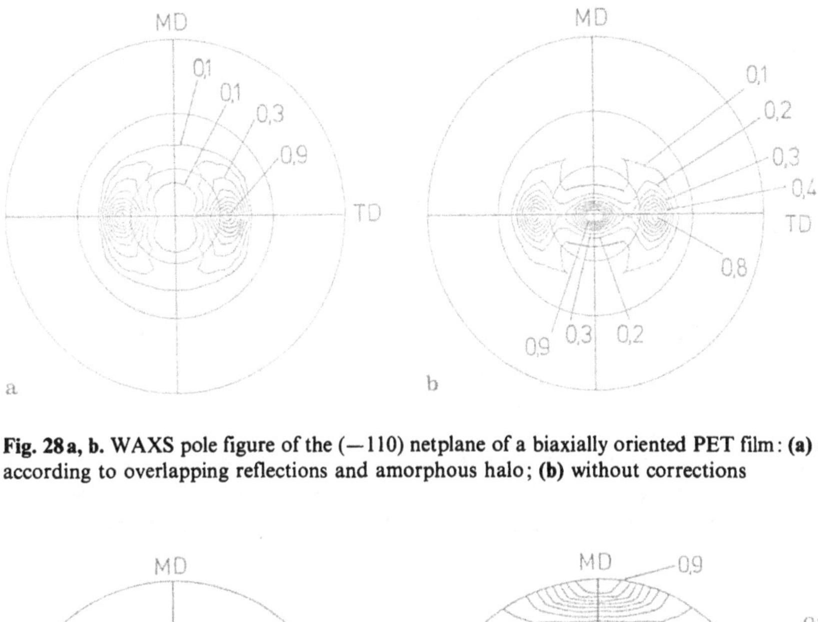

Fig. 28a, b. WAXS pole figure of the (—110) netplane of a biaxially oriented PET film: **(a)** corrected according to overlapping reflections and amorphous halo; **(b)** without corrections

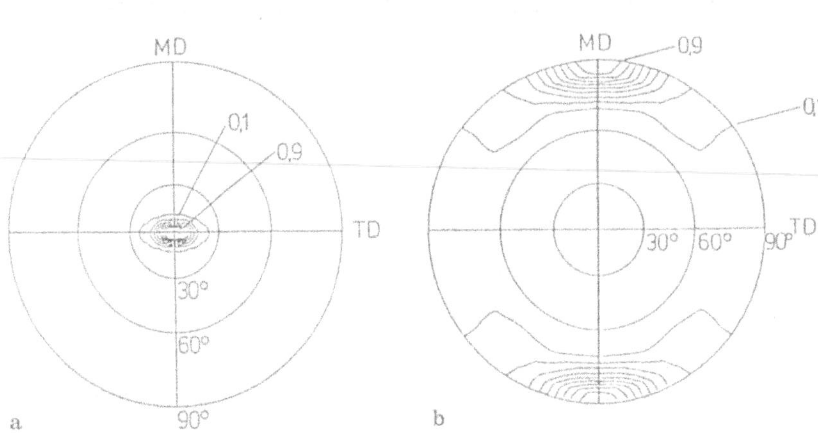

Fig. 29a, b. WAXS pole figures of two different netplanes measured on a biaxially oriented PET film: **(a)** (100) netplanes, **(b)** (—105) netplanes

In the case of PET the (100) and (—105) net planes are of special interest. The normals of the (100) planes are perpendicular to the planes of the benzene rings, while the normals of the (—105) planes are nearly parallel to the chain direction. The corrected pole figures of the two net planes are presented in Fig. 29. From these figures it can be seen immediately, that in the case of the investigated biaxially oriented sample the benzene rings are mostly aligned parallel to the film surface (see Fig. 29a), while the chains exhibit a relatively broad distribution around the first drawing direction, MD (see Fig. 29b).

3.6.2 Small Angle X-Ray Pole Figures

Similar to the WAXS pole figures, with SAXS one can obtain the orientation distribution of the normals onto the lamella surfaces. The measurements are based on the same instrumental technique, but with conventional X-ray sources the acquisition of one single SAXS pattern needs as much as several hours or days. Therefore the acquisition of a complete SAXS pole figure would require a tremendous amount of time and has never been carried out. The availability of S.R. reduces the required time to a few hours.

Röber et al. [56] determined the SAXS pole figures of uniaxially oriented PET. Figure 30 shows the results obtained with a PET film which was drawn in the MD direction at a temperature of $T_d = 40$ °C and was afterwards crystallized at $T_c = 240$ °C. After the orientation process, which took place by necking, the optical birefringence was $\Delta n_0 = 0.190$, indicating nearly the highest orientation attainable. On the left side of Fig. 30, SAXS curves are plotted for $\varphi = 90°$ and different values of α in the range from $\alpha = 90°$ to $\alpha = 30°$, i.e. the SAXS diagram for a given α reflects the periodicity of the lamellae, which are oriented with their surface normals pointing out of the sample film, lying in the plane defined by the film normal and the MD direction (because of $\varphi = 90°$) and enclosing the angle $90° - \alpha$ with the film surface. The scattering curves become broader with decreasing α, while the maximum shifts to larger angles. This indicates, that the more the lamella normals are inclined relative to the film surface, the smaller is the mean lamella distance and

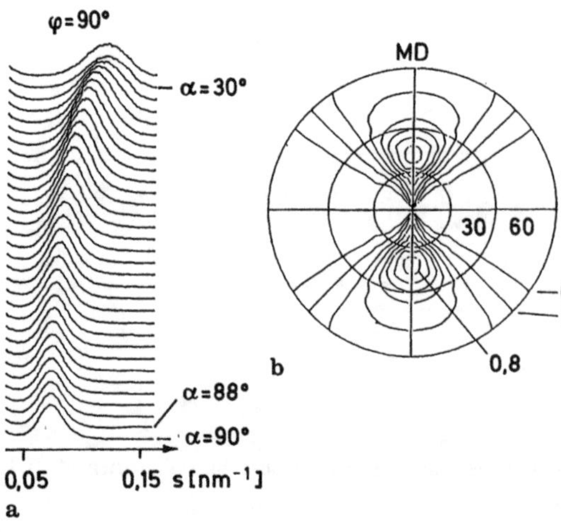

Fig. 30a. Small-angle X-ray scattering curves of a uniaxially cold drawn PET film ($T_d = 40$ °C, $T_c = 240$ °C, $\Delta n_0 = 0.19$) obtained for $\varphi = 90°$ and various fixed angles α; **(b)** SAXS pole figure of the same sample (MD = draw direction)

the more is the regular packing of the lamellae distorted. The pole figure on the right side of Fig. 30 is obtained from the integrated intensities of the SAXS curves. It illustrates the orientation distribution of the lamellae normals without taking the lamellae distance into account. From the pole figure it can be recognized, that the normals to the crystal lamellae are mainly lying in the plane defined by the MD direction and the normal to the film surface, with a preferred orientation at $\alpha = 45°$.

A significantly different result is found, if the drawing in the MD direction is performed at a temperature above the glass transition temperature, for example at $T_d = 92$ °C. Then the material is deformed homogeneously resulting in a lower degree of orientation, as is indicated by the birefringence of $\Delta n_0 = 0.111$. The corresponding pole figure is given in Fig. 31. In this case the normals of the lamellae are mainly pointing directly to the draw direction, MD.

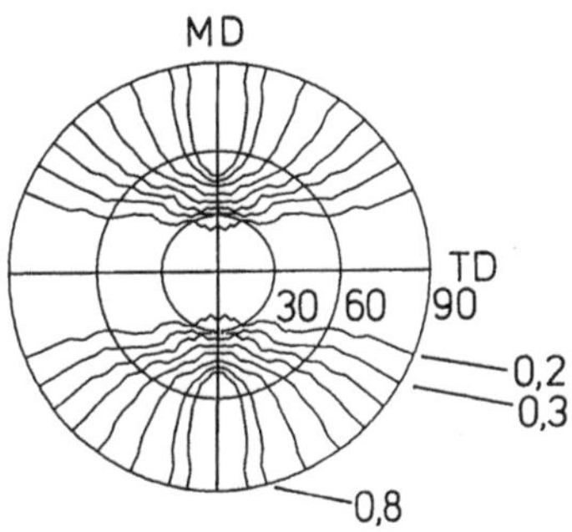

Fig. 31. SAXS pole figure of a PET film drawn uniaxially at $T_d = 92$ °C ($T_c = 240$ °C, $\Delta n_0 = 0.111$)

3.7 Deformation Studies

3.7.1 SBS-Block Copolymers

The structural changes in styrene-butadiene-styrene (SBS) block copolymers during deformation have been studied by Polizzi et al. [57]. Aliphatic or aromatic oil was added to the samples. The material forms regularly ordered regions of cyclindrical shape, consisting of polystyrene, which are embedded in a matrix of polybutadiene. The added oil mingles with the polybutadiene phase. By means of SAXS the dimensions and distances of the polystyrene cylinders can be determined. The samples were successively stretched to different draw ratios and the SAXS patterns were registered by a vidicon detector, simultaneously. The time resolution of the measurements was 10 s, which is sufficiently short compared to the relaxation time of the material of about one minute. Figure 32a shows the scattering pattern of a sample after drawing it to the fourfold of the initial length. At the left side of the primary beam stop two reflections emerge, which are confined to layer lines. From the reflections the mean distance L, the height h, and the diameter 2r of the polystyrene cylinders was calculated. The results are plotted in Figs. 32b, 32c and 32d, respectively, as a function of the draw ratio λ, for three different sample materials. One can see, how the cylinders are deformed due to the elongation of the sample.

3.7.2 Crazing

Many polymers show a formation of crazes, when mechanical stress is applied. These are crack-like defects containing fibrils, which extend from one end of the crack to the other. The formation and propagation of crazes is of great importance for the technical application of polymer materials, since they are often the reason why polymers fail and fatigue. Crazes can be observed by electron microscopy, but no time resolved measurements during application of stress can be performed with this technique. The large electron density difference between the fibrils and the voids causes a strong SAXS. Thus it is possible to follow the formation and propagation of crazes by means of time resolved SAXS measurements utilizing S.R.

This was done by Rothwell et al. [58] and Brown et al. [59–61]. The materials under investigation were mainly polystyrene, polycarbonate, and polymethyl methacrylate. Step-wise or oscillatory deformations were performed and the influence of thermal treatment or chemical additives, such as solvents and plasticizers was studied.

a

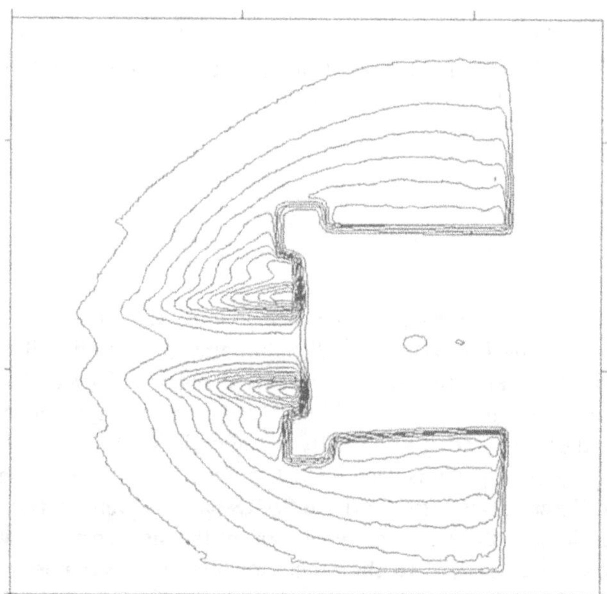

Fig. 32. **(a)** Two-dimensional small-angle X-ray scattering pattern of an SBS block copolymer with 33% aliphatic oil content and stretched to a draw ratio $\lambda = 4$. **(b)** Long period L in SBS block copolymers as a function of the draw ratio λ (pure SBS and SBS with different oil contents). **(c)** height of the polystyrene cylinders h in SBS block copolymers as a function of the draw ratio λ (pure SBS and SBS with different oil contents). **(d)** diameter of the polystyrene cylinders 2r in SBS block copolymers as a function of the draw ratio λ (pure SBS and SBS with different oil contents)

The high brightness of S.R. makes it possible to measure the scattering of a single isolated crack [60]. Brown et al. irradiated a crack of 4.5 mm length and some µm thickness in polystyrene with a highly collimated X-ray beam of 50 µm width. Figure 33 shows the observed SAXS profile at different positions along the crack, the increasing numbers indicate positions which are closer to the crack tip. A detailed analysis of the scattering curves leads to the determination of the crack volume and of the diameter of the fibrils.

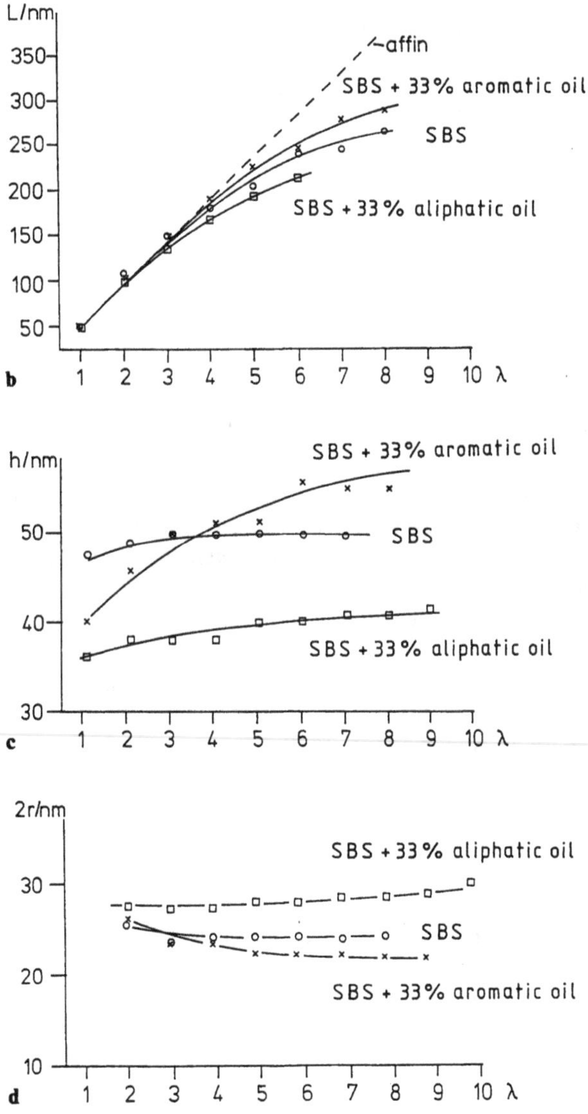

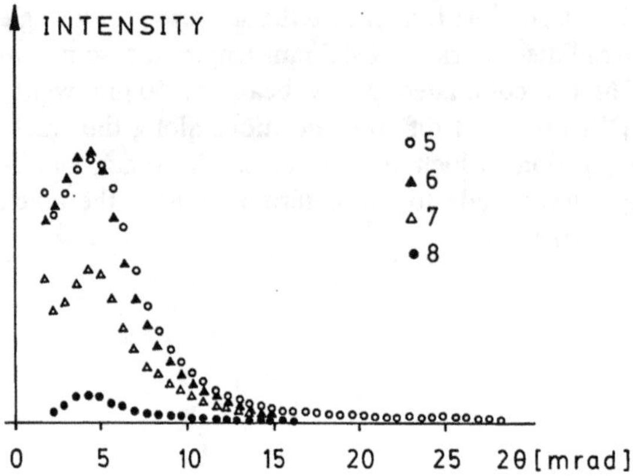

Fig. 33. Small-angle X-ray scattering of a single crack in plasticized polystyrene measured at different positions along the crack. A smaller number at the profile designates a position closer to the crack tip [60]

3.8 Phase Separation

3.8.1 Polymer Blends

Most polymers are not arbitrarily miscible with each other. Often a critical temperature exists, above which a phase separation into two mixtures of different compositions occurs. For mixtures of low molecular weight polymers the opposite effect can also be observed. Two incompatible polymers mix with each other, if a certain temperature is exceeded. An example for such a system are mixtures of polystyrene and polybutadiene.

An important question concerning phase separation processes is, whether a homogeneous polymer mixture exhibits critical behaviour as the binodal is approached. The density fluctuations within a polymer mixture lead to SAXS. Close to the critical temperature these density fluctuations will change significantly, thus influencing the SAXS in a characteristic manner. In the homogeneous state the scattering intensity I is given by an Orstein-Zernicke function

$$I(q) = \frac{I(0)}{1 + \xi^2(T, \varphi)\, q^2} \tag{7}$$

where $\xi(T, \varphi)$ is the correlation length of the mixture as a function of the temperature T and of the volume fraction φ of one polymer. Application of the classical mean field theory leads to an expression of the form

$$\xi(T) = \xi_0 \frac{(T - T_s)^\nu}{T_s^\nu} \tag{8}$$

where T_s is the spinodal temperature, ξ_0 the critical correlation length, and v is the critical exponent. According to the mean field theory ξ_0 is equal to $v^{1/3}$, with v being the monomer volume, while v should be equal to 0.5.

Russell [62] studied the SAXS of a 50% mixture of polystyrene with molecular weight M = 2000 and polybutadiene with M = 1000. Figure 34 shows the scattering profiles obtained at different temperatures. At the highest temperature, the mixture is in the homogeneous state, with decreasing temperature it approaches the binodal and at last a phase separation takes place, which is manifested in an increase of the scattering intensity. The analysis of this measurements according to Eqs. (7) and (8) is presented in Fig. 35. The correlation length ξ, as determined according to Eq. (7) from the SAXS data, is plotted versus $(T - T_s)/T_s$ according to Eq. (8). The

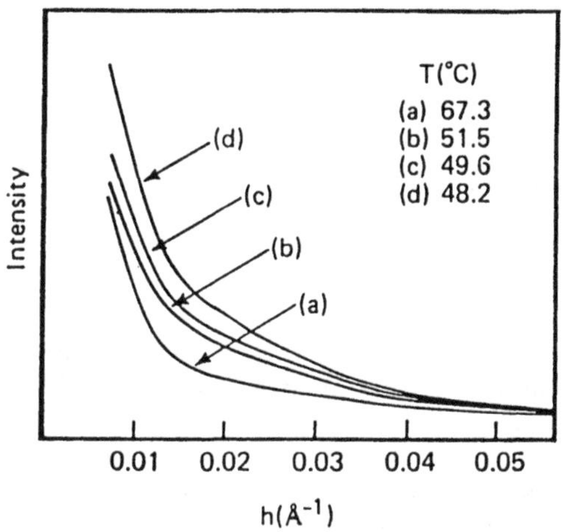

Fig. 34. Small-angle X-ray scattering profile of a mixture of 50% polystyrene (M = 2000) and 50% polybutadiene (M = 1000) at the different temperatures indicated

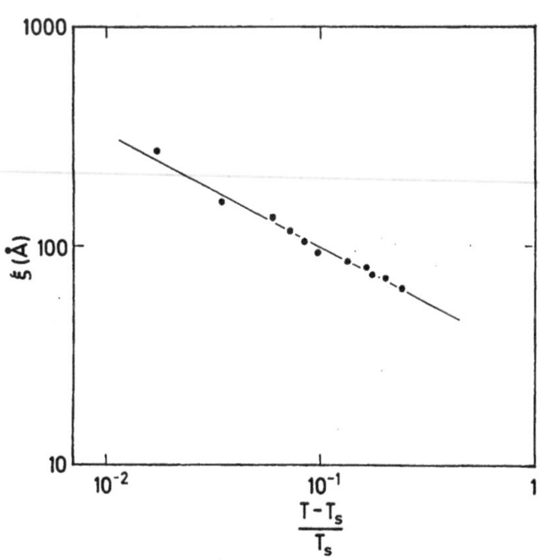

Fig. 35. Dependence of the correlation length ξ on the reduced temperature $(T - T_s)/T_s$ of a mixture of 50% polystyrene (M = 2000) and 50% polybutadiene (M = 1000)

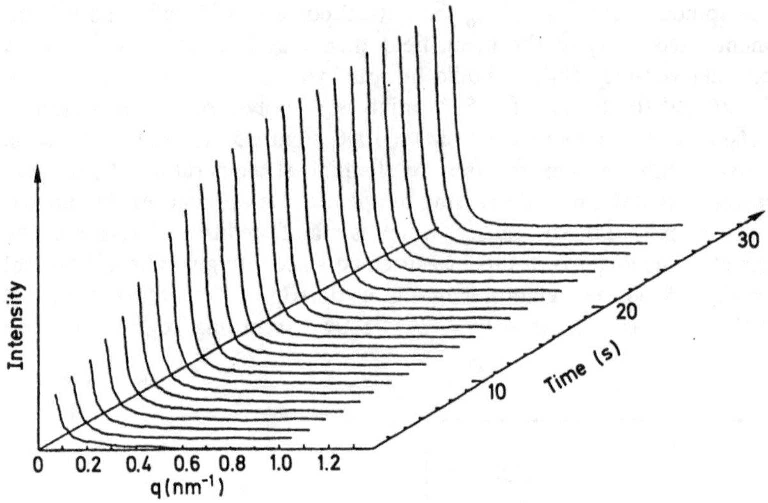

Fig. 36. Temporal development of the small-angle X-ray scattering profile of a mixture of 60% polystyrene (M = 2000) and 40% polybutadiene (M = 1000) after quenching to below the cloud point curve (108 °C)

behaviour is proved to be in a good agreement with the mean field arguments. The value found for the critical correlation length from the plot of Fig. 35 is $\xi_0 = 0.34$ nm, the critical exponent is found to be $\nu = 0.507$.

Russell et al. also investigated the kinetics of this phase separation [62]. The homogeneous mixture was rapidly cooled to a temperature within the spinodal region. The subsequent change of SAXS was then followed by means of S.R. and a photodiode array detector. The result is given in Fig. 36. According to the theory of Cahn and Hilliard [63–65], the temporal development of the SAXS during a spinodal decomposition is given by

$$I(q, t) = I(q, 0)\, e^{-2R(q)t}, \tag{9}$$

where R(q) is the amplification factor which is defined by thermodynamical quantities,

$$R(q) = -M\left(\frac{\partial^2 f}{\partial c^2}\right) q^2 - 2MKq^4 \tag{10}$$

M being the molecular mobility, f(c) the free energy density of the mixture as a function of the concentration, and K a gradient energy term. In the case of polymer mixtures Eq. (10) becomes [66, 67]

$$R(q) = -q^2 \Lambda(q) \left\{ \frac{1}{(N_A + N_B)\,\Phi_A \Phi_B} - 2\chi + \frac{(aq)^2}{36\Phi_A\Phi_B} \right\} \tag{11}$$

with N_i and Φ_i being the molecular weight and the volume fraction of the i-th component, respectively, χ is the Flory-Huggins interaction parameter, a is the statistical segment length, and Λ is the Onsager coefficient.

In Fig. 37 the intensity data corresponding to different s-values within Fig. 36 are plotted logarithmically as a function of time. For short times the course of the intensity is obviously well described by Eq. (9). From the slope of the curves in

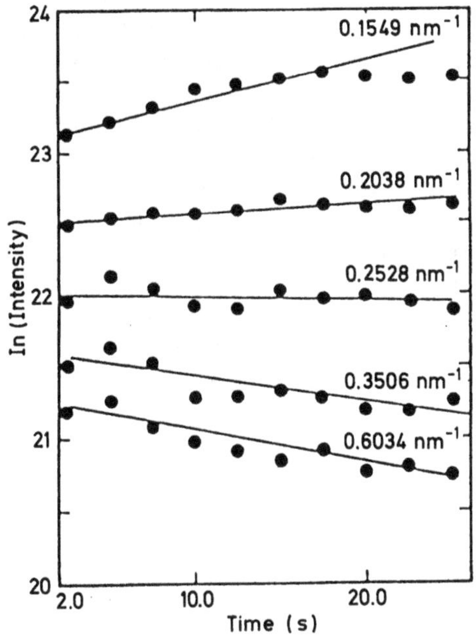

Fig. 37. Logarithm of the intensity versus time obtained from the data of Fig. 36 for various fixed scattering angles q given as a parameter to each curve

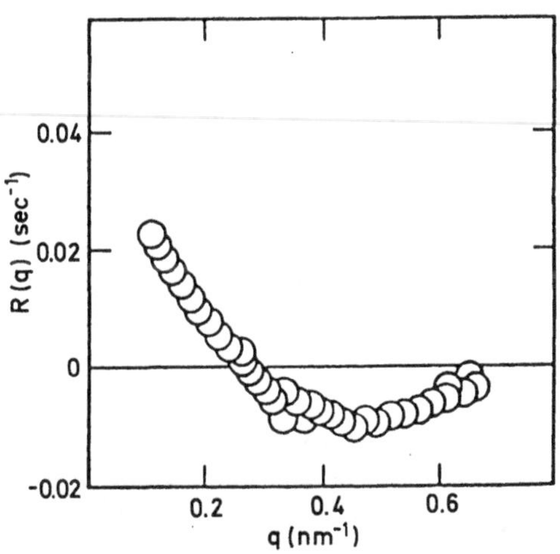

Fig. 38. Amplification factor R(q) obtained as a function of q from the slopes of the curves in Fig. 37

Fig. 37 the amplification factor R(q) can be determined. The result is shown in Fig. 38. At last it shall be mentioned, that the value $q = q_s$ for which $R(q_s) = 0$ in Fig. 38, can be put into Eq. (11) in order to calculate the value of the interaction parameter χ.

3.8.2 Block Copolymers

Block copolymers, which are composed of incompatible blocks will separate into ordered phases, leading to a pronounced SAXS. The formation of ordered cylindrical structures in SBS copolymers has already been mentioned earlier in this text (see Sect. 3.7.1).

Figure 39 reports the SAXS measurements on SBS block copolymers at different temperatures around the order/disorder transition temperature [68]. At low temperatures the copolymer is in the ordered state, while it disorders with increasing temperature. In the figure the measured data are represented by points. Leibler described the scattering curves by a theoretical approach, which besides the molecular weight of the blocks, only contains the binary interaction parameter between segments of different blocks as an adjustable parameter [69]. The solid curves in Fig. 39 are fitted according to this theory by an appropriate choice of the interaction parameter, showing an excellent agreement of theory and experimental data.

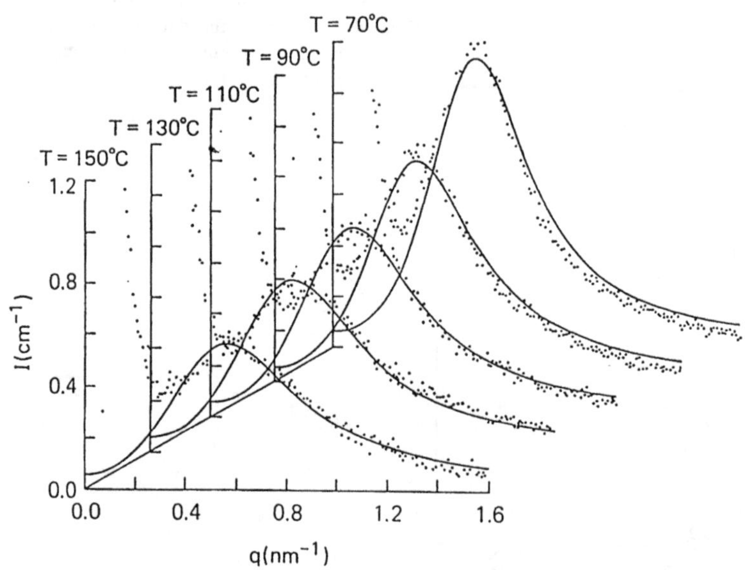

Fig. 39. Small-angle X-ray scattering profiles of a polystyrene/polybutadiene block copolymer (molecular weight of the blocks M = 18600) at different temperatures T above the order/disorder transition temperature. The *points* indicate the measured intensity values while the *solid lines* are obtained by fitting a one-parametric theoretical model

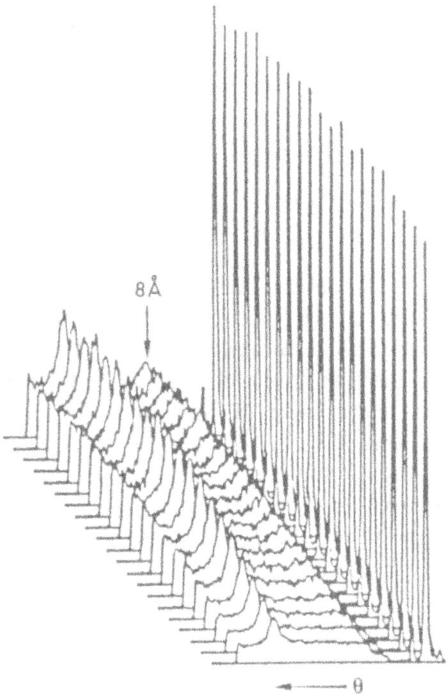

Fig. 40. Change of the wide-angle X-ray scattering pattern of *cis*-polyacetylene during a reaction with iodine vapour at room temperature. The progression of the reaction is indicated by the appearance of a characteristic 0.8 nm reflection. The strong peak at the right side is caused by a semitransparent beam stop

3.9 Chemical Reactions and Aggregations

3.9.1 Jodine in Polyacetylene

Riekel [70] studied the reaction of *cis*-polyacetylene with jodine vapour. Time resolved measurements of WAXS were performed, while jodine was guided onto the polyacethylene sample in an especially designed reaction cell. The result is given in Fig. 40. During the reaction the appearence of a new reflection can be observed, which corresponds to a repetition distance of 0.8 nm. The absorption of X-rays increases significantly during the progress of the reaction, an effect which was corrected by means of a semitransparent beam stop. The intensity that has passed this beam stop can be seen in the strong peak on the right side of each single WAXS pattern. All the patterns have been normalized to the same integral intensity within this peaks in order to correct the absorption.

From the change of the intensity of the characteristic 0.8 nm reflection, conclusions about the kinetics of the reaction can be drawn. It was found, that the reaction, which is characterized by a half time of 340 s, is not a diffusion controlled process.

3.9.2 Gelation

The gel formation in acrylamide/biacrylamide copolyesters was investigated by Giessler et al. [72]. The cross linking reactions used to form the gels, produces two phase structures, which consist of a homogeneous solution of polymer chain segments and of fractal structures with sizes between 10 nm and 100 nm and dimensionalities between 0 and 3, depending upon the degree of cross linking.

4 Other Methods Employing Synchrotron Radiation

4.1 EXAFS

With some polymers which contain atoms of higher atomic numbers, measurements of the Extended X-Ray Absorption Fine Structure (EXAFS) and of the X-Ray Absorption Near Edge Structure (XANES) are feasible. By determining the absorption as a function of the photon energy close above an absorption edge of these atoms, information about the structure in the vicinity of this atoms can be yielded. One has to be aware of the fact, that the energies of the absorption edges shift towards lower values with decreasing atomic number. This implies, that for elements like sulphur, X-ray absorption measurements are possible in principle, but problems arise due to the high absorption of the low energetic photons. Nevertheless, recently EXAFS measurements at the K-absorption edge of sulphur in natural rubber have been reported [71].

As an example of EXAFS studies in polymer science, the work of Pan et al. should be mentioned, they investigated, among other things, nickel atoms in nafion membranes [73]. The perfluorosulfonic acid ionomer was neutralized with Ni^{2+} cathions. The results of the measurements indicate, that the Ni^{2+} ions are in an octahedral site with six oxygen atoms as nearest neighbours. The degree of disorder in the Ni—O distances was found to the comparable to that in ionic crystals of the same material. Both findings are valid in dry as well as in water soaked membranes. But it has been observed, that an addition of water leads to an enhanced contribution of the second shell of neighbours to the EXAFS signal. This second shell is formed by adjacent Ni^{2+} ions. This indicates, that the addition of water enhances the local ordering of the cathion environment.

4.2 Fluorescence Depolarization

The time structure of S.R. can be used for time correlated single photon counting experiments in the low energy region of the spectrum, i.e. from visible light to UV. The short light pulse caused by a passing bunch has a typical length of several hundred picoseconds, the exact value differs for each storage ring. The light pulse might perform a certain excitation of the molecules. If a single fluorescent photon is emitted as a consequence of this excitation before the occurrence of the next flash, the time correlation to the exciting photon is established. Thus curves can be obtained, showing the number of emitted photons as a function of the elapsed time after the excitation pulse. One advantage of S.R., compared to laser sources, is the well defined pulse shape, which facilitates the deconvolution of excitation pulse and response. Another advantage is the easy tunability of the excitation wavelength. However, concerning the brightness and the shortest attainable pulse lengths, lasers are the superior sources.

In the field of polymer science, a time correlated single photon counting technique has been applied to polymer solutions and solid polymers in order to measure the depolarization of fluorescence light as a function of time after excitation. This depolarization is caused by moving chromophores and thus gives information about the mobility of the polymer chains [74]. Usually fluorescent labels have to be added to the material. This can be done simply by mixing long and stiff labels with the polymer or by incorporating the label chemically into some of the chain molecules. In the experiments the excitation photons are vertically polarized and the fluorescence photons, which are registered perpendicular to the incident beam exhibit a certain degree of polarization, too. If the chromophore gradually changes its orientation due to molecular motion this polarization is increasingly lost for an increasing time between excitation and emission. An appropriate measure for the degree of polarization is the fluorescence anisotropy

$$r(t) = \frac{I_V(t) - I_H(t)}{I_V(t) + 2I_H(t)} \tag{12}$$

where $I_V(t)$ and $I_H(t)$ are the measured intensities of the vertically and horizontally polarized fluorescence light, respectively, as a function of the time t after the excitation pulse. The anisotropy r(t) is proportional to the second moment $M_2(t)$ of the orientation autocorrelation function of the emission transition moment of the fluorescent label, i.e. the relation

$$r(t) = r_0 \frac{M_2(t)}{M_2(0)} \tag{13}$$

holds, where r_0 is the fundamental anisotropy, which is a time independent molecular parameter. $M_2(t)$ can be calculated for different models of motion and the results obtained can be compared with the experimental data [75].

As an example, some results obtained by Viovy et al. on bulk polybutadiene are presented [76, 77]. 1% of the chains were labelled by anthracene in their middle, as is shown in Fig. 41. The evolution of the anisotropy decay with temperature is given in Fig. 42. As expected, the decay rate increases as the temperature rises, indicating an increase of the chain mobility in the vicinity of the chromophore.

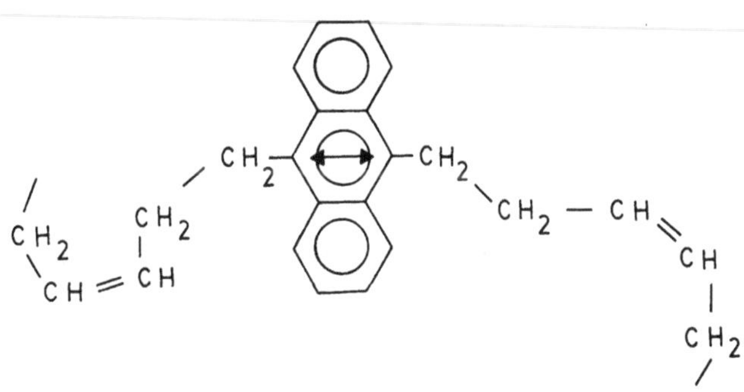

Fig. 41. Polybutadiene with an incorporated anthracene label

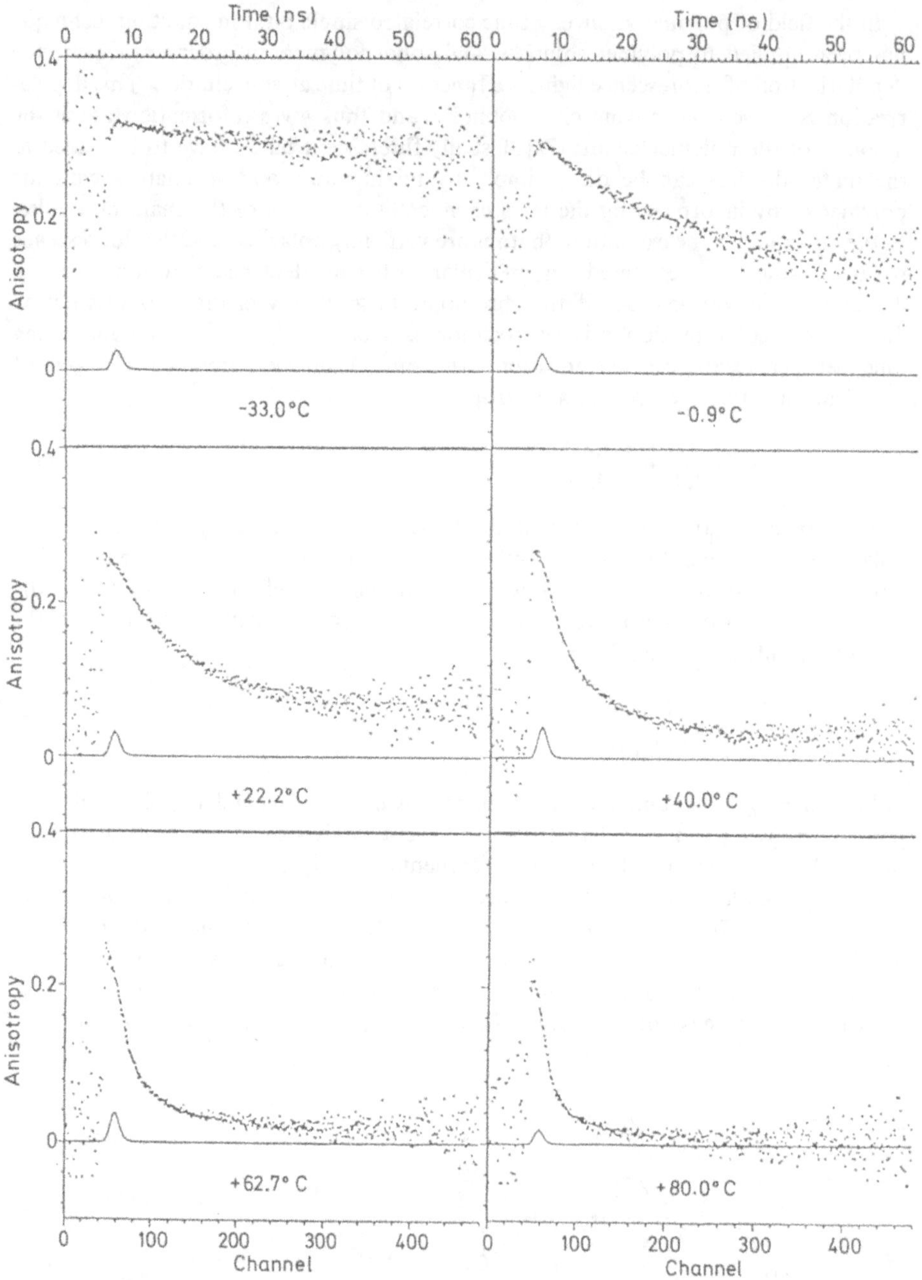

Fig. 42. Fluorescence anisotropy decay of bulk polybutadiene labelled with anthracene in the middle of the chains observed at various temperatures

5 Future Prospects

Recent developments in the field of the application of synchrotron radiation in polymer research concentrate on the improvement of time resolved SAXS instrumentation. One aim is a further increase of the photon flux in order to be able to measure on samples, which are even weaker scatterers. Since polymer materials often exhibit relatively broad X-ray reflections, an increase of the energetic bandwidth of the monochromatized primary beam often would mean no significant disadvantage concerning the resolution of the experiment, but would bring the advantage of a higher intensity. Therefore one way to obtain higher flux is to increase the bandpass of the monochromator. In this field the attention has been focussed onto the development of synthetic multilayer structures. Stephenson [78], for example, reports about experiences with a high intensity multilayer monochromator consisting of 200 layer pairs of tungsten and silicon with a d-spacing of 2.35 nm. The energy resolution of this monochromator was found to be $\Delta E/E = 1.1 \cdot 10^{-2}$ FWHM at 6 keV, thus being about two orders of magnitude lower than with perfect crystal monochromators. The reflectivity at the given setup was about 26%.

Another way to reach higher photon flux is to install SAXS instruments at wiggler or undulator beamlines [79].

The extremely high brightness of undulators will also facilitate the design of SAXS instruments with higher limits of resolution because a better collimation can be achieved without losing too much intensity. The resolution of the focussing instruments will likely also be increased by improving the quality of the X-ray optics, especially that of the imaging mirrors.

In the field of detector development, devices will have to be designed, which are able to cope with high counting rates as they are expected with future high flux beam lines and especially with fast time resolved measurements in the subsecond region. Photodiode arrays and charge coupled devices are basically favourable. However, the limited overall size and readout rate limit the application of this detectors in many cases. But it can be expected, that these two parameters will be improved in the future. In some special cases, the recently developed imaging plates [80] could have some future importance, too. They represent a considerable improvement over photographic film and are in some applications competitive to photo diode arrays. However, fast time resolved measurements are not possible with this type of detectors.

By improving the instruments, a further increasing number of investigations in polymer research will be realizable. This includes time resolved measurements down to the millisecond regime, as well as the examination of the structures of thin layers (e.g. Langmuir Blodgett type films), surfaces and boundaries between different phases.

Surfaces structures especially will be of future interest. They can be investigated by glancing angle X-ray diffraction, since with this technique the X-rays penetrate only through a few atomic layers at the sample surface. Therefore, the measured reflections can give information about two-dimensionally ordered surface structures.

6 References

1. Elsner G, Riekel C, Zachmann HG (1985) Adv. in Polym. Sci. 67: 1
2. Guinier A (1963) X-ray diffraction, Freeman, San Francisco
3. Glatter O, Kratzky O (eds) (1982) Small angle X-ray scattering, Academic, New York
4. Vainshtein BK (1966) Diffraction of X-rays by chain molecules, Elsevier, London
5. Bramer VR (1974) Colloid & Polym. Sci. 252: 504
6. Blundell DJ (1970) Acta Cryst. A26: 472
7. Vonk CG, Kortleve G (1967) Kolloid Z. & Z. Polym. 220: 19
8. Vonk CG, Kortleve G (1968) Kolloid Z. & Z. Polym. 225: 124
9. Strobl GR (1973) J. Appl. Cryst. 6: 365
10. Strobl GR, Schneider M (1980) J. Polym. Sci., Polym. Phys. Ed. 18: 1343
11. Ruland W (1971) J. Appl. Cryst. 4: 70
12. Vonk CG (1973) J. Appl. Cryst. 6: 81
13. Blundell DJ (1978) Polymer 19: 1258
14. Koberstein JT, Morra B, Stein RS (1980) J. Appl. Cryst. 13: 34
15. Wunderlich B (1976) Macromolecular physics, Academic, New York
16. Alexander LE (1961) X-ray diffraction methods in polymer science, Wiley, New York
17. Bonse U, Hart M (1966) Z. Phys. 189: 151
18. Bonse U, Pahl R, Nusshardt R (1987) Acta Cryst. A43: 259
19. Bonse U, Pahl R, Nusshardt R (1987) HASYLAB annual report, p 383
20. Bordas J, Randall JT (1978) J. Appl. Cryst. 11: 434
21. Bordas J, Munro JH, Glazer AM (1976) Nature 262: 541
22. Stuhrmann H (1988) Top. Curr. Chem. 145: 151
23. Gabriel A (1977) Rev. Sci. Instr. 48: 1303
24. Hendrix J, Fuerst H, Hartfiel B, Dainton D (1982) Nucl. Instr. Meth. 201: 139
25. Hendrix J, Lentfer A (1986) Nucl. Instr. Meth. A252: 246
26. Prieske W, Riekel C, Koch MHJ, Zachmann HG (1983) Nucl. Instr. Meth. 208: 435
27. Radeka V (1984) Nucl. Instr. Meth. 226: 209
28. Livingston WC, Harvey J, Slaughter J, Trumbo D (1976) J. Appl. Optics 15: 40
29. Barbe DF (1975) Proc. IEEE 63: 38
30. Hoffmann JD, Lauritzen JI, Passaglia E, Ross GS, Frolen LJ, Weeks JJ (1969): Kolloid Z. & Z. Polym. 231: 564
31. Barham PJ, Chivers RA, Keller A, Martinez-Salazar J, Organ SJ (1985) J. Mat. Sci. 20: 1625
32. Martinez-Salazar J, Barham PJ, Keller A (1985) J. Mat. Sci. 20: 1616
33. Elsner G, Koch MHJ, Bordas J, Zachmann HG (1981) Makromol. Chem. 182: 1263
34. Zachmann HG, Gehrke R (1986) In: Sedlacek B (ed) Advances in indirect methods of polymer morphology characterization, de Gruyter, Berlin, p 119
35. Gehrke R, Riekel C, Zachmann HG Polymer, submitted
36. Gehrke R (1985) PhD-Thesis, University of Hamburg
37. Elsner G, Zachmann HG, Milch JR (1981) Makromol. Chem. 182: 657
38. Schober H, Zachmann HG (1986) HASYLAB annual report, p 321
39. Zachmann HG, Wiswe D, Gehrke R, Riekel C (1985) Makromol. Chem., Suppl. 12: 175
40. Schultz JM, Lin JS, Hendricks RW, Petermann J, Fohil RM (1981) J. Polym. Sci., Polym. Phys. Ed. 19: 609
41. Prieske W, to be published
42. Fan Q, Stamm M, Zietz R, Fischer EW (1987) HASYLAB annual report, p 277
43. Grubb DT, Liu JJH, Caffrey M, Bilderback DH (1984) J. Polym. Sci., Polym. Phys. Ed. 22: 367
44. Fronk W, Heise B, Neppert B, Schubach HR, Wilke W (1984) Colloid & Polym. Sci. 262: 99
45. Koch MHJ, Bordas J, Schoela E, Broecker HC (1979) Polym. Bull. 1: 709
46. Stach W, Riekel C, Zietz R, Holland-Moritz K, Broecker C, Koch MHJ (1982) HASYLAB annual report, p 167
47. Caffrey M, Bilderback DH (1983) Nucl. Instr. Meth. 208: 495
48. Holl WW, Heise B (1983) HASYLAB annual report, p 209
49. Buchner S, Zachmann HG (1987) HASYLAB annual report, p 285
50. Riekel C (1983) Makromol. Chem., Rapid Commun. 4: 479

51. Schouterden P, Riekel C, Koch MHJ, Groeninckx G, Reynaers H (1985) Polym. Bull. 13: 533
52. Russell TP, Koberstein JT (1985) J. Polym. Sci., Polym. Phys. Ed. 23: 1109
53. Koberstein JT, Russell TP (1986) Macromolecules 19: 714
54. Bark M, Zachmann HG (1987) HASYLAB annual report, p 391
55. Röber S, Gehrke R, Zachmann HG (1987) Mat. Res. Soc. Symp. Proc. 79: 205
56. Röber S, Boesecke P, Zachmann HG (1988) Makromol. Chem., Makromol. Symp. 15: 295
57. Polizzi S, Boesecke P, Zachmann HG (1987) HASYLAB annual report, p 282
58. Rothwell WS, Martinson RH, Gorman RL (1983) Appl. Phys. Lett. 42: 422
59. Brown HR, Sindoni Y, Kramer EJ, Mills PJ (1984) Polym. Eng. & Sci. 24: 825
60. Brown HR, Mills PJ, Kramer EJ (1985) J. Polym. Sci., Polym. Phys. Ed. 23: 1857
61. Brown HR, Njoku NG (1986) J. Polym. Sci., Polym. Phys. Ed. 24: 11
62. Russell TP, Hadziioannou G, Warburton W (1985) Macromolecules 18: 78
63. Cahn JW (1968) Trans. Metall. Soc. AIME 242: 166
64. Cahn JW, Hilliard JE (1958) J. Chem. Phys. 28: 258
65. Cahn JW, Hilliard JE (1959) J. Chem. Phys. 31: 668
66. deGennes PG (1980) J. Chem. Phys. 72: 4756
67. Pincus P (1981) J. Chem. Phys. 75: 1966
68. Owens JN, Gancarz K, Koberstein JT, Russell TP (1986) Bull. Am. Phys. Soc. 31: 519
69. Leiber L (1980) Macromolecules 13: 1602
70. Riekel C, Menke K (1984) Mol. Cryst. Liquid Cryst. 105: 245
71. Hormes J, Kuetgens U, Weymans G: Report at the spring meeting of the German Physical Society (DPG), Hamburg, 1988
72. Hecht AM, Geissler E (1987) Macromolecules 20: 2485
73. Pan HK, Yarusso DJ, Knapp GS, Cooper SL (1983) J. Polym. Sci., Polym. Phys. Ed. 21: 1389
74. Viovy JL, Monnerie L (1985) Adv. Polym. Sci. 67: 99
75. Viovy JL, Monnerie L, Brochon JC (1983) Macromolecules 16: 1845
76. Viovy JL, Monnerie L, Merola F(1985) Macromolecules 18: 1130
77. Viovy JL, Frank CW, Monnerie L (1985) Macromolecules 18: 2606
78. Stephenson GB (1988) Nucl. Instr. Meth. A266: 447
79. Riekel C (1987) In: Foundation Phase Report, ESRF, Grenole
80. Miyahara J, Takahashi K, Amemiya Y, Kamiya N, Satu T (1986) Nucl. Instr. Meth. A246: 572

Use of Synchrotron Radiation to Study State-Selected Ion-Molecule Reactions

Paul-Marie Guyon[1] and Eric A. Gislason[2]

[1] Also, Laboratoire des Collisions Atomiques et Moleculaires (associated to CNRS), Bât. 351, Université Paris-Sud, 91405 Orsay Cedex, France

[2] Also, Department of Chemistry, University of Illinois at Chicago, Chicago, Illinois 60680, U.S.A.

Table of Contents

The threshold photoelectron-photoion coincidence technique using synchrotron radiation has been employed to study state-selected ion-molecule reactions. We review the experimental procedure and discuss our results on five representative systems: $[N_2Ar]^+$; $[N_2H_2]^+$; $[ArH_2]^+$; $[HeH_2]^+$; and $[H_2H_2]^+$. Recent theoretical work on these systems is also summarized.

LURE (CNRS, CEA, and Université Paris-Sud), Orsay, France

1 Introduction

The study of ion-molecule reactions using state-selected reagents has become a very exciting area of molecular dynamics. We have developed an experimental apparatus in Orsay which utilizes the properties of our tunable synchrotron radiation source at LURE to prepare ions in selected vibronic levels and then to study their reactions. The ions are state-selected using the TPEPICO (threshold-photoelectron/photoion coincidence) method [1].

The apparatus has been used to study a variety of ion-molecule collision processes. These include the simple charge-transfer reaction [2]

$$N_2^+(X, A; v) + Ar \rightarrow N_2 + Ar^+ . \quad (1)$$

Here the N_2^+ ion is prepared in vibrational level v of either the $X^2\Sigma_g^+$ or $A\,^2\Pi_u$ electronic state. This reaction of state-selected N_2^+ ions has been the subject of many other experimental [3, 4] and theoretical [5–8] studies. We have also measured charge-transfer cross sections for the reaction [9, 10]

$$N_2^+(X, A; v) + H_2 \rightarrow N_2 + H_2^+ . \quad (2)$$

For this system at low energies the major product channel is chemical reaction to give $N_2H^+ + H$, and it is interesting to study the competition between the two product channels. At the same time we measured charge transfer cross sections for $Ar^+(^2P_{3/2}, ^2P_{1/2})$ colliding with H_2. The third type of ion-molecule system we have studied is [11]

$$H_2^+(v) + He \rightarrow HeH^+ + H \quad (3a)$$

$$\rightarrow He + H^+ + H . \quad (3b)$$

In this case both chemical reaction (3a) and collision-induced dissociation (3b), denoted CID, can occur. During the course of this work we developed a theoretical method for distinguishing the two (nearly) degenerate electronic product states produced in CID [12, 13]. This type of analysis gives considerable insight into the dissociation process. The extensive literature on H_2^+ + He collisions is reviewed in Refs. 11–13. Finally, we have studied charge-transfer in the symmetric system [14]

$$H_2^+(v) + H_2 \rightarrow H_2 + H_2^+ , \quad (4)$$

and we have also measured cross sections for CID [15]. This system has been widely studied, both experimentally and theoretically. Recent representative papers are those of Ng [16], DePristo [17], and Schatz [18]. Earlier work is reviewed in these references.

Section 2 of this paper describes the apparatus we use at Orsay to study ion-molecule collision processes. The following four sections discuss the results we have obtained for reactions (1)–(4). These are a representative sample of the type of systems we can study at present. Section 7 discusses future prospects for using synchrotron radiation to study other ion-molecule reactions.

2 Experimental Method Used at Orsay

2.1 Apparatus

The experimental technique is an extension of the threshold photoelectron-photoion coincidence technique (TPEPICO) as used in Orsay for the study of the unimolecular dissociation of vibronically-selected molecular ions, and has been previously described [1]. Briefly, synchrotron radiation from ACO, the Orsay electron storage ring, is dispersed by a 1 m McPherson normal incidence monochromator, equipped with a 2400 1/mm holographic grating. It is then refocused into the photoionization chamber of a dual time-of-flight spectrometer (Fig. 1). This chamber contains the reactant ion's parent gas. In the present investigations the wavelength resolution was set at 1.0 Å (FWHM), corresponding to an energy resolution of 20 meV at hν ≅ 15 eV. All photoelectrons are accelerated over 0.7 cm by a 2 V/cm electric field and detection is restricted to threshold electrons; i.e., those having nominally zero energy within a 20 meV band pass, on the basis of temporal and angular discrimination.

Parent ions of internal energy equal to the nominal photon energy or their ionic reaction products are selectively detected in delayed coincidence with threshold electrons. For this purpose the threshold electron signal is used to trigger an electric pulse producing a 10 V/cm electric field for 15 μs; i.e., longer than the total ion TOF. The primary ions are thus accelerated over 0.3 cm of the ionization chamber and refocused by a 3 element electrostatic immersion lens onto an effusive jet of

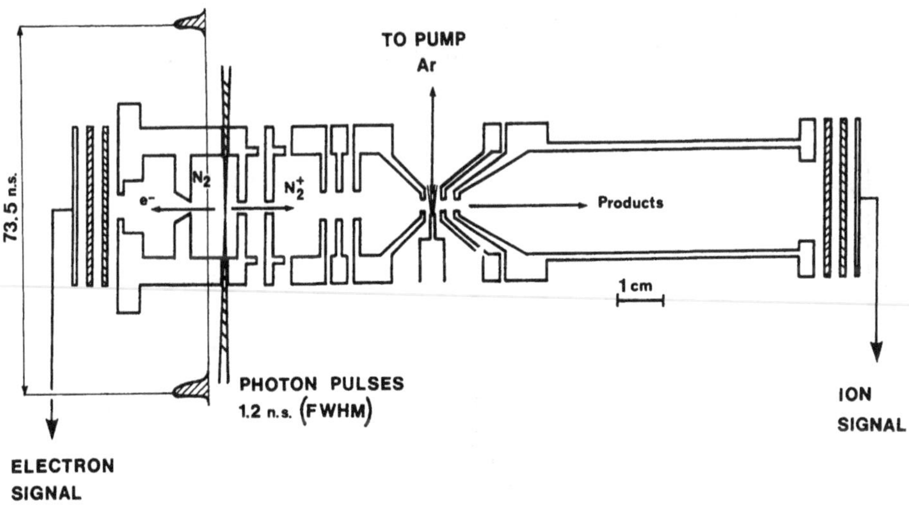

Fig. 1. Scale drawing of the dual electron-ion time-of-flight spectrometer. Dispersed synchrotron radiation passes through the chamber containing the N_2 reactant gas. The photoelectrons are accelerated to the left detector, and the N_2^+ ions are accelerated to the right toward the Ar beam. Unreacted N_2^+ ions and Ar^+ products are then accelerated through the TOF chamber to the right detector

neutral target gas. The latter emerges from a 0.4 mm ID hypodermic needle and is directed at the center of the collision region defined by two diaphragms covered by gold mesh. They are spaced by 0.3 cm and present circular apertures of 0.2 cm and 0.3 cm ID, respectively. A constant 20 V/cm electric field is established between these diaphragms in order to extract the low energy product ions. The median potential defines, within a halfwidth of 3 eV, the average laboratory collision energy.

The vacuum chamber containing the dual TOF spectrometer is evacuated by a 2000 l/s turbomolecular pump. Operating pressures in the ionization chamber and in the collision region are estimated at 2×10^{-3} and 3×10^{-3} Torr for a background pressure of 4×10^{-5} Torr in the vacuum chamber.

2.2 Experiment Procedure

Figure 2 shows the threshold photoelectron spectrum (TPES) of N_2 obtained at the resolution used in the present experiment. The levels v = 0 to 4 of N_2^+ ($X^2\Sigma_g^+$) and v = 0 to 6 of $N_2^+(A^2\Pi_u)$ are easily distinguishable. The spectrum contrasts with the usual HeI PES which shows negligible excitation of v > 1 for the X state because of unfavorable Franck-Condon factors. Here the higher levels are observed because of autoionization. The wide tunability of synchrotron radiation offers the possibility, as the wavelength is being scanned, of excitation to the frequently rich spectrum of autoionising neutral states, whose subsequent decay allows the formation of ion levels inaccessible by direct transitions.

Figure 3 illustrates a typical coincidence spectrum. It was obtained for reaction between N_2^+(X, v = 2) and Ar at a collision energy of 8 eV. The abscissa shows the time elapsed between the detection of the threshold electron and the arrival of the ion at the detector. The ordinate shows the number of coincidences per channel obtained during a 6.7 min. accumulation time. The relative cross section for

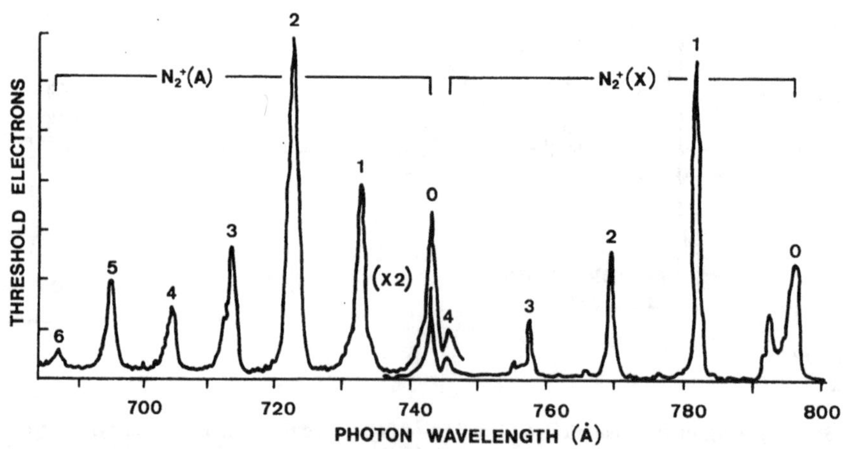

Fig. 2. Threshold photoelectron spectrum of N_2 [2] obtained under the experimental conditions described in the text. The count rate for the N_2^+ (X, v = 1) peak was 2000 counts/sec. It is possible to increase the resolution and cleanly separate peaks for N_2^+ (X, v = 4) and N_2^+ (A, v = 0) [9]

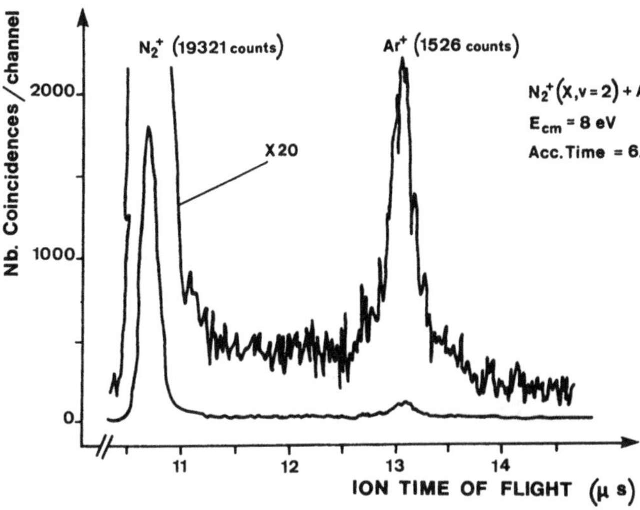

Fig. 3. Typical time-of-flight spectrum [2] showing fast unreacted N_2^+ ions and slow Ar^+ product ions resulting from the reaction of N_2^+ (X, v = 2) with Ar at $E_{cm} = 8$ eV

charge transfer is obtained from the ratio of the secondary Ar^+ peak area to the sum of the primary N_2^+ and the secondary Ar^+ peaks. It is not possible to obtain absolute cross sections, because the target gas density is not known. In some cases, the relative cross sections can be put on an absolute scale by normalizing to absolute cross sections obtained by other investigators on the same system. Alternatively, one can run simultaneous experiments with two different reactant ions scattered by the same target gas. If absolute cross sections are known for one of the ions, the results for the other ion can be normalized to those values. We have used both procedures in our work.

The background in Fig. 3 is due to false coincidences. These are produced when an ion, uncorrelated with a detected threshold electron, generates a stop signal. One notices that, at variance with most coincidence experiments, the background is structured with maxima underlying the 2 peaks of interest. This is due to the pulsing method used to extract the primary ions from the source region. In order to correct for this we generate a false coincidence spectrum using random start pulses and subtract the latter from the former after normalization to the total number of starts. Recently, we have been able to greatly reduce the rate of false coincidences by using a new pulsing method (see Sect. 7).

3 Charge Transfer in the N_2^+ + Ar System

We have measured charge transfer cross sections [2] for state-selected N_2^+(X, A; v) ions colliding with Ar. The experiments give relative cross sections; they are put on an absolute basis by normalizing to the Ar^+ + Ar charge transfer cross section. It is also necessary to correct the cross sections measured for the N_2^+ (A; v) states for the finite lifetime of these radiative states. The details of these adjustments

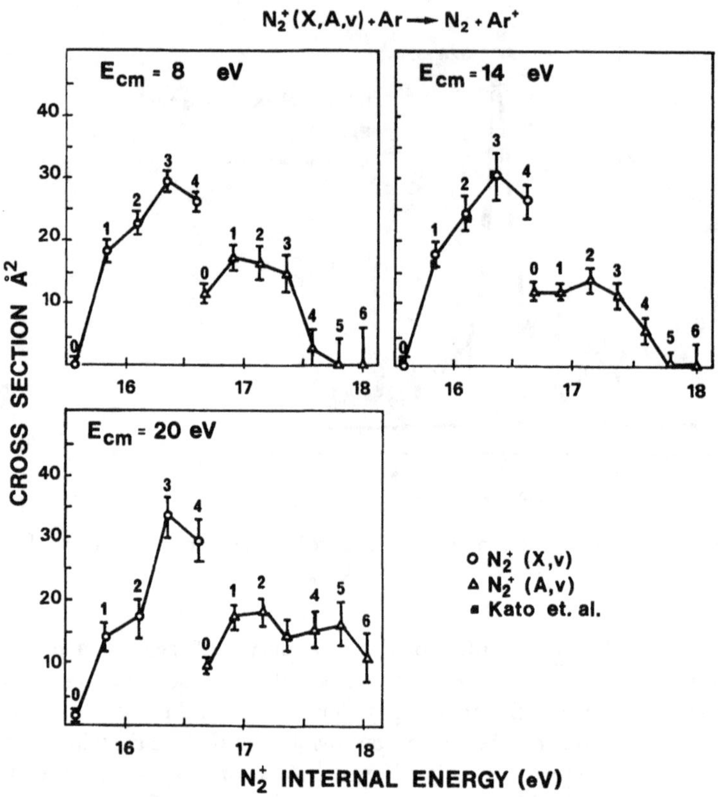

Fig. 4. Absolute charge transfer cross sections in Å^2 for three collision energies plotted against the N_2^+ internal energy [2]. (The zero of energy corresponds to ground state neutral N_2 + Ar). The data for N_2^+ (X, v) states are shown as *circles* and the data for N_2^+ (A, v) as *triangles*. The *small numbers* above each data point indicate the reactant vibrational level. The *black squares* in the middle block show Kato's cross sections [3] at a collision energy of 11.8 eV adjusted to our absolute values at 14 eV

are given in [2]. The absolute cross sections are shown in Fig. 4 for the relative collision energies 8, 14, and 20 eV.

Certain features of the results are quite interesting. The cross sections show a strong dependence on the vibrational quantum number for both reactant electronic states. If the Franck-Condon principle were valid for the nonadiabatic transitions which occur in this system, then the charge transfer cross section would be independent of the reactant vibrational level [19]. It is well known that the Franck-Condon principle breaks down badly at low collision energies for most charge transfer systems. The most remarkable result seen in Fig. 4 is the very small cross section for $N_2^+(X; v = 0)$ + Ar at all three collision energies; its maximum value is 1.6 Å^2 at 20 eV. (By comparison the cross sections for other $N_2^+(X; v)$ + Ar states are at least 14 Å^2.) This occurs even though there is a product state, $Ar^+(^2P_{3/2}) + N_2(v = 0)$, which is only 0.18 eV away thus, this. In addition, the Franck-Condon factor for the transition $N_2^+(X; v = 0) \rightarrow N_2$ (v = 0) is 0.92;

transition is strongly "allowed". A third notable feature of the results is the smaller reactivity of the $N_2^+(A; v)$ states. For example, the vibronic states $N_2^+(A; v = 0) + Ar$ and $N_2^+(X; v = 4) + Ar$ differ in energy by only 0.06 eV, but their charge transfer cross sections at 20 eV differ by a factor of two.

The results obtained in our laboratory as well as by other experimentalists [3, 4] have inspired a considerable amount of theoretical work on this system [2, 5–8]. Archirel and Levy [7] have calculated a set of potential energy surfaces for the states $N_2^+(X) + Ar$, $N_2^+(A) + Ar$, and $N_2 + Ar^+(^2P)$ as well as the couplings between these surfaces using a novel computational technique. From their results they developed a set of *diabatic* vibronic potential energy curves, and they assumed that transitions could occur when two curves crossed. Cross sections were computed using either the Demkov or Landau-Zener formula, as appropriate, and good agreement was obtained with the experimental values in most cases. Nikitin et al. [8] have taken a somewhat similar approach to this system. They estimated the *adiabatic* vibronic interaction curves for this system, and they assumed that transitions

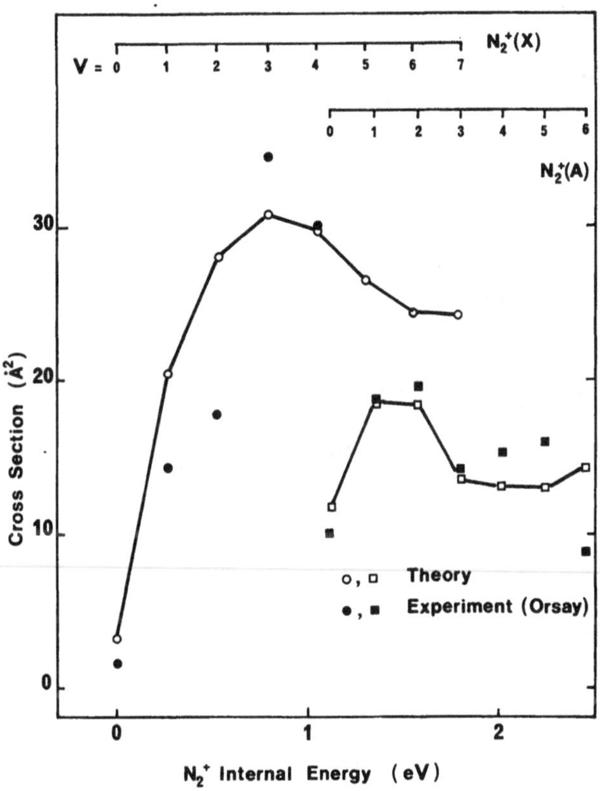

Fig. 5. Charge transfer cross sections for N_2^+ (X, A; v) + Ar at $E_{cm} = 20$ eV plotted against the internal energy of the N_2^+ ion. (The zero of energy corresponds to N_2^+ (X; v = 0) + Ar.) Theoretical results for v = 0 to 7 of the X state and v = 0 to 6 of the A state are shown as *open circles* and *squares*, respectively [6]. Experimental results [2] are shown as *solid circles* and *squares*. The lines drawn through the theoretical values are only a guide for the eye

could occur when any two curves have an avoided crossing. The Landau-Zener formula was used to estimate the cross sections. Their calculated cross sections for $N_2^+(X; v = 0 - 3) + Ar$ agree well with our experiments.

The most detailed calculations on this system have been carried out by Spalburg et al. [5] and by Parlant and Gislason [6]. The latter work used the potential energy surfaces of Archirel and Levy [7] in a classical path treatment of the charge transfer process. Here the translational motion is treated classically, and the vibronic degrees of freedom are treated quantum-mechanically using the time-dependent Schrodinger equation. This type of calculation is expected to be quite accurate at energies above 5 eV. The theoretical cross sections at 20 eV are shown in Fig. 5. It is seen that the agreement between experiment and theory is very good for both reactant electronic states.

In addition to reproducing the absolute cross sections, the theoretical models discussed here [5–8] can give explanations of the various experimental results discussed above. For example, the strong dependence of the cross sections on the vibrational level of the N_2^+ ion is not surprising once the vibronic curves [5, 7, 8] for this system are examined. In particular, the adiabatic vibronic state which goes asymptotically to $N_2^+(X; v = 0) + Ar$ does not cross another vibronic state at any value of R; this explains why the charge transfer cross section is so small in low energy collisions. The smaller reactivity of the $N_2^+(A)$ state shows up nicely in Fig. 5 based upon the calculations of Parlant and Gislason [6]. They attribute this to two causes. First, the electronic coupling between the states $N_2^+(A) + Ar$ and $N_2 + Ar^+(^2P)$ is considerably smaller than between $N_2^+(X) + Ar$ and $N_2 + Ar^+(^2P)$. In addition, one third of all $N_2^+(A) + Ar$ collisions take place on the $\Omega = 3/2$ electronic surface. In this case charge transfer can only occur to the $\Omega = 3/2$ state of $Ar^+(^2P_{3/2}) + N_2$; charge transfer to $Ar^+(^2P_{1/2}) + N_2$ cannot occur. This significantly reduces the cross sections for the $\Omega = 3/2$ reactants.

4 Charge Transfer in the $N_2^+ + H_2$ and $Ar^+ + H_2$ Systems

We have recently measured state-selected charge transfer cross sections [9, 10] for collisions of $N_2^+(X, A; v) + H_2$ as well as for the system

$$Ar^+(^2P_{3/2}, {}^2P_{1/2}) + H_2 \rightarrow Ar + H_2^+ . \quad (5)$$

The cross sections have been put on an absolute basis by normalizing to earlier non-state-selected measurements. The results are shown in Fig. 6 for the two laboratory energies studied, 10.5 and 14.5 eV. This corresponds to relative energies of 0.70 and 0.97 eV for $N_2^+ + H_2$ and 0.50 and 0.69 eV for $Ar^+ + H_2$. It is seen that the charge transfer cross sections for $Ar^+(^2P_{1/2}) + H_2$ are much larger than for the other reactant states. These, in turn, are much smaller than values obtained for the $N_2^+ + Ar$ system (see Fig. 4). It should be noted that the ionization potentials of Ar (15.76 eV) and H_2 (15.43 eV) are similar, so the cross section differences cannot be attributed to this factor.

Figure 6 also shows cross sections measured in other laboratories. The data of Ng and coworkers [20] agree well with our results at both collision energies. By

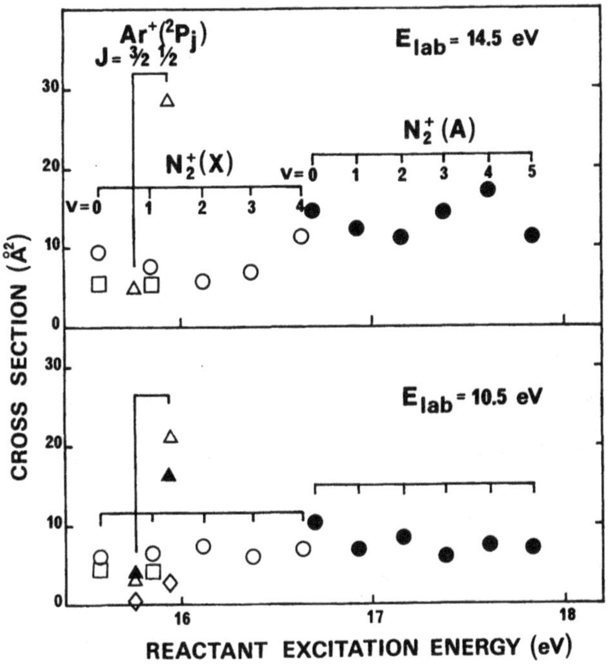

Fig. 6. Absolute charge transfer cross sections [9] for state-selected N_2^+ (X, v = 0 to 4), N_2^+ (A, v = 0 to 5), and $Ar^+(^2P_{3/2}, ^2P_{1/2})$ colliding with H_2 at laboratory collision energies of 10.5 eV (*lower panel*) and 14.5 eV (*upper panel*). Our results for N_2^+(X), N_2^+(A), and $Ar^+(^2P_J)$ are shown as *open circles*, *closed circles*, and *open triangles*, respectively. Ng's results [20] for N_2^+ and Ar^+ are shown as *open squares* and *closed triangles*, respectively. The data of Tanaka et al. [21] for Ar^+ are represented by *open diamonds*

comparison the cross sections measured by Tanaka et al. [21] for $Ar^+(^2P_J) + H_2$ are much smaller than those obtained by us and by Ng. A possible explanation of this difference is that the H_2^+ product ions produced in Tanaka's experiments go on to react with H_2 to form H_3^+ molecules. This would greatly reduce the apparent cross section for H_2^+.

The explanation of the relatively small charge transfer cross sections was given in the Introduction. For both N_2^+ and Ar^+ colliding with H_2 the major product channel at these collision energies is chemical reaction to give N_2H^+ or ArH^+. We expect that collisions with small impact parameters will give chemical reaction, so charge transfer can only occur in grazing collisions. Seen in this light, the large cross sections for $Ar^+(^2P_{1/2}) + H_2$ in Fig. 6 require an explanation. This state is nearly resonant with $Ar + H_2^+(v = 2)$. If H_2 and H_2^+ are both in the J = 1 rotational level, the energy difference is only 0.0083 eV. We have shown [9] that the Demkov model predicts a cross section in good agreement with the measured value.

For the other reactant states the competition between reaction and charge transfer can be described by a simple model similar to that of North and Leventhal [22]. We restrict the discussion to collisions of $N_2^+(X; v) + H_2$. The model assumes that all collisions with impact parameters less than b_R give chemical reaction. Then there is a second range of impact parameters, $b_R \leqq b \leqq b_C$, where charge transfer can occur to give $N_2 + H_2^+$. The estimated cross sections for reaction and charge transfer are

$$Q_R = \pi b_R^2 \tag{6a}$$

$$Q_{CT} = \frac{1}{2}\pi(b_C^2 - b_R^2)\,. \tag{6b}$$

The factor of $\frac{1}{2}$ comes from the fact that for near-resonant charge transfer 50% of the collisions remain in the reactant channel. If chemical reaction were impossible, the charge transfer cross section would be $\frac{1}{2}\pi b_C^2$. The model requires independent estimates of b_R and b_C. At 1 eV the reactive cross section was determined by Hierl et al. [23] to be 20 Å^2; this gives $b_R = 2.52$ Å. We estimate the parameter b_C from measurements of the charge transfer cross sections for $N_2^+ + Ar$ (see Sect. 3). Typical cross sections in Fig. 4 for this system at the same relative *velocity* as in the $N_2^+ + H_2$ experiments are 20 ± 5 Å^2. If these numbers are substituted into Eq. (6b), the predicted cross sections are $Q_{CT} = 10 \pm 5$ Å^2 at 1 eV. A similar analysis predicts $Q_{CT} = 5 \pm 5$ Å^2 at a relative energy of 0.70 eV. (The charge transfer cross section grows with energy because Q_R is declining with energy.) These theoretical predictions are in good agreement with the cross sections shown in Fig. 6. We conclude that this simple model describes well the competition between reaction and charge transfer in the $N_2^+ + H_2$ system.

5 Collisions Between H_2^+ and He

We have studied [11] collisions of state-selected $H_2^+(v)$ ions with He to give (see Eq. 3) $HeH^+ + H$ (chemical reaction) and $He + H^+ + H$ (collision-induced dissociation — CID) at a relative collision energy of 3.1 eV. Individual vibrational levels between 0 and 6 were studied. The cross sections were put on an absolute scale by normalizing to the earlier work of Chupka and coworkers [24]. Both product channels are endothermic for $H_2^+(v = 0)$ ions, chemical reaction by 0.81 eV and CID by 2.65 eV. Thus, one expects vibrational excitation of the reactant ions to greatly increase the cross sections for both products. This is confirmed by our results. Absolute cross sections for HeH^+ products are shown in Fig. 7 and for the H^+ products in Fig. 8.

Examination of Fig. 7 shows that the reactive cross section rises monotonically for v = 0 to 5 but levels off for v = 6. The dependence of the cross section on v agrees well with the earlier work of Chupka [24]. The agreement with the data of Turner et al. [25] is less satisfactory; their cross sections level off at v = 3 and 4 at values considerably below our results. We attribute the difference to the manner in which they prepare their ions and estimate the distribution of vibrational levels in their beam. By comparison the TPEPICO technique prepares ions in a single vibrational level. The theoretical calculations of Whitton and Kuntz [26] and of

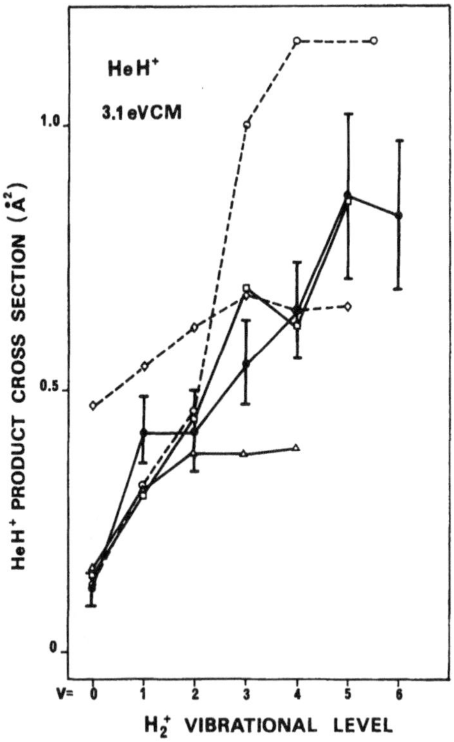

Fig. 7. Absolute total cross sections as a function of the vibrational level v for production of HeH^+ in collisions of $H_2^+(v)$ + He at $E_{cm} = 3.1 \pm 0.7$ eV. Experimental data: *solid circles*, our results [11]; *solid squares*, Chupka [24]; *solid triangles*, Turner [25]. Theoretical cross sections: *diamonds*, Whitton and Kuntz [26]; *open circles*, Joseph and Sathyamurthy [27]

Joseph and Sathyamurthy [27] are in qualitative agreement with the experiments, but the absolute cross sections are not well reproduced.

In the course of our experiments we also measured the distribution for the velocity component of the product HeH^+ ions along the initial relative velocity vector. This distribution varies rapidly with the vibrational levels of the H_2^+ reactant ions. For $v < 3$ the HeH^+ ions are predominantly scattered backward (with respect to the He relative motion), but for $v > 3$ the products are scattered mainly forward. Clearly, the reaction does not go through a long-lived complex. We believe the results can be explained as follows. Reactive collisions for $v < 3$ require small impact parameter collisions to overcome the reaction endothermicity. This type of collision gives products scattered backwards. By comparison, the chemical reaction is quite exothermic for $v > 3$. Consequently, reaction can occur in large impact parameter collisions giving forward scattered products. Since large impact parameters are intrinsically more probable, the angular distribution moves forward as v increases. We also note that the simple spectator stripping model [28] does not work for any reactant vibrational level.

The cross sections for CID in Fig. 8 show, as expected, a strong dependence on vibrational level, increasing from 0.045 Å² for $v = 0$ to 1.10 Å² for $v = 6$. For the most part the cross sections measured by Chupka [24] agree with ours, but their values for $v = 0$ and 1 are smaller than our results. A possible explanation is that their apparatus discriminated against H^+ ions produced at large scattering angles.

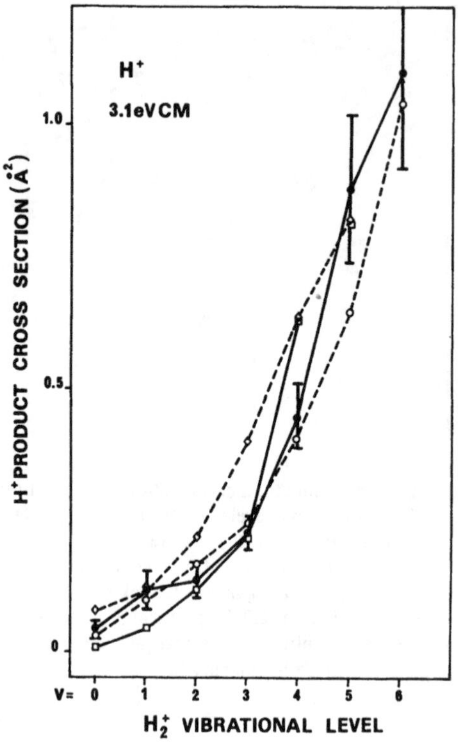

Fig. 8. Absolute total cross sections as a function of the vibrational level v for production of H^+ in collisions of H_2 (v) + He at $E_{cm} = 3.1 \pm 0.7$ eV. Experimental data: *solid circles*, our results [11]; *solid squares*, Chupka [24]. Theoretical cross sections: *diamonds*, Whitton and Kuntz [26]; *open circles*, Schneider et al. [29]

Presumably CID from v = 0 or 1 of H_2^+ requires a very strong collision, which gives product H^+ scattered to large angles. The theoretical cross sections of Whitton and Kuntz [26] as well as of Schneider et al. [29] are also shown in Fig. 8. The agreement with our experiments is fairly good, although the theoretical cross sections are too large for v = 2 and 3.

An important aspect of CID of homonuclear diatomic ions such as H_2^+ is that there are two low-lying product electronic states. These correspond to $H_A^+ + H_B$ + He and to $H_A + H_B^+$ + He, where we have labeled the protons to distinguish them. In principle, these two states are degenerate when the three particles are infinitely separated. Recently, we have shown [12], however, that the presence of the He atom breaks the degeneracy of the two product states, even at very large separation of all three particles. This allows the two product electronic states to be distinguished at the end of the collision according to the following rule: If the final relative velocity between the H^+ and He is greater than between the H and He, the system is in the excited electronic state; if the reverse is true, the system is in the ground state. Recent trajectory-surface hopping (TSH) calculations in our laboratory [13] using the two lowest potential energy surfaces for the HeH_2^+ system have shown that a significant fraction of the CID products are formed on the excited potential energy surface even at low collision energies.

In our experiments we also measured the velocity component along the initial relative velocity vector of the product H^+ ion. The results showed that all H^+ ions produced from H_2^+ in levels v = 0 and 1 were appearing on the excited electronic surface. (The results for H^+ ions produced from H_2^+ in levels v = 2 to 6 were ambiguous; that is, the product electronic state could not be uniquely identified from the measured velocity component.) This was completely unexpected. We are presently studying this process theoretically using the TSH method, and we plan to repeat the measurements in the near future. It is safe to conclude, however, that a full understanding of the HeH_2^+ system requires a consideration of both electronic surfaces.

6 Collisions Between H_2^+ or D_2^+ and H_2

The $H_2^+ + H_2$ system represents a considerable increase in complexity when compared to $H_2^+ + He$. The number of internal degrees of freedom rises from 3 to 6, and as many as 8 potential energy surfaces could be involved in the reaction [30]. The product channels which must be considered are symmetric charge transfer (CT) to give $H_2 + H_2^+$, atom (AT) or proton (PT) transfer to give $H_3^+ + H$, and collision-induced dissociation (CID) to yield $H_2 + H^+ + H$. This complexity suggests that it will be difficult to obtain sufficient experimental information to rigorously test the various theoretical models which have been applied to this system. Fortunately, the cross sections for these processes vary with collision energy in very different ways. As in the case of $N_2^+ + H_2$, discussed earlier, the chemical reaction channels AT and PT dominate at energies below 2 eV. At higher energies CT is the primary process, but CID is also a significant product channel.

This system, which is of great importance in astrophysics, has been extensively studied at high collision energy [31–33], but few experiments have been performed with state-selected H_2^+ ions. Anderson et al. [34] studied CT, AT, and PT at low collision energy as a function of reactant vibrational level, and Campbell et al. [35] performed the first PEPICO experiment on this system. They measured relative CT cross sections at energies above 4 eV. Although these data suffered from low statistics, they showed a significant dependence of the cross section on vibrational energy. The results could not be reproduced by the available theoretical models. These results prompted us to repeat the experiment with our TPEPICO method determining cross sections for all four product channels.

Cross sections were measured [14, 15] for $H_2^+(v)$ and $D_2^+(v)$ reactants colliding with H_2 for v = 0 to 10 at cm translational energies of 4, 8, 12, and 16 eV. In $H_2^+ - H_2$ collisions the product ions produced by CT have the same mass as the reactant ions but a different kinetic energy. They were distinguished from the primary ions by their time of flight (TOF) in our apparatus. A typical TOF spectrum is shown in Fig. 9. It is seen that the primary products are H^+ from CID and H_2^+ from CT. The absence of any H_3^+ peak shows that the reactive AT and PT channels have negligible cross sections at 8 eV. (This was true at all of the collision energies studied.) Absolute cross sections for CT and CID as a function of v are shown in Fig. 10 for the four collision energies. The data were normalized to the absolute cross sections obtained by Ng and coworkers [16, 36]. The interesting variation of the CT cross sections with vibrational energy, as well as their weak dependence on collision energy, is now well understood, thanks to the theoretical work of DePristo and

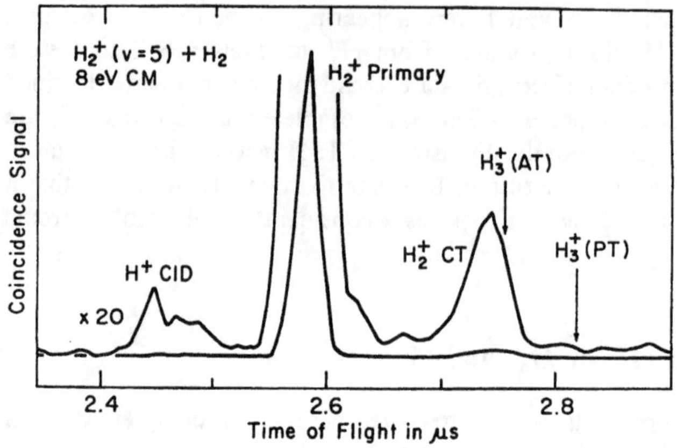

Fig. 9. A typical (unsmoothed with no background substracted) time-of-flight spectrum for the products of the H_2^+ (v = 5) + H_2 reaction at E_{cm} = 8 eV [15]. The collection time was 20 minutes. The *arrows* indicate the predicted times for H_3^+ produced by atom-transfer (AT) and proton-transfer (PT). The other product peaks are labeled in the Figure

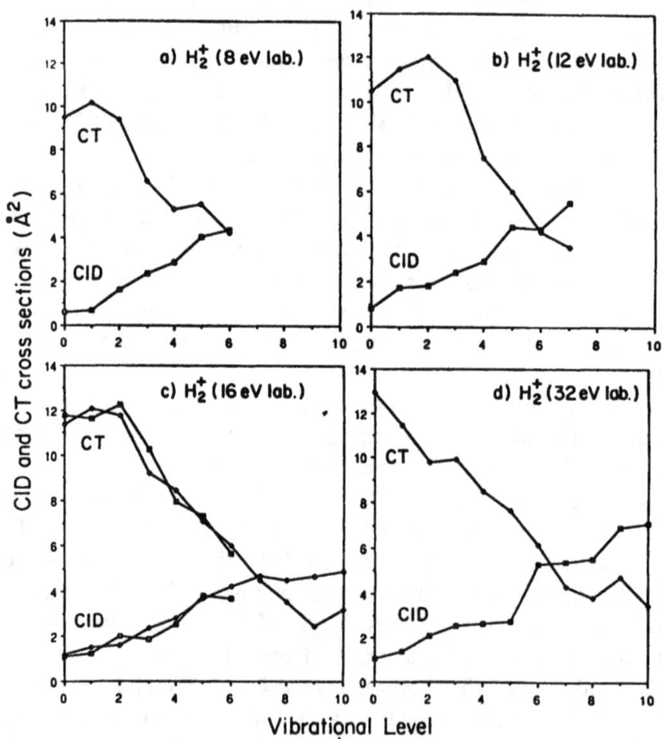

Fig. 10. Absolute cross sections in Å^2 plotted against the H_2^+ vibration level v for charge transfer (CT) and collision-induced dissociation (CID) for collisions of $H_2^+(v)$ + H_2. [15] The laboratory collision energies are given on each panel; the center of mass energies are half as great. The cross sections were normalized to the absolute values reported by Lee et al. [36] at 16 eV lab; those cross sections for v = 0–6 are also shown in panel (*c*)

coworkers [17, 37]. The CT process is dominated at low v by resonant CT at relatively large $H_2 - H_2$ separation R ($\approx$4 a.u.). However, as v increases the CT becomes less energy-resonant, due to the unfavorable Franck-Condon factors for near-resonant transitions. Thus, the transition occurs at smaller values of R, and the cross section decreases accordingly.

Anderson et al. [34] have measured cross sections for CID of state-selected $D_2^+(v)$ + HD. Unfortunately, no data exists for $H_2^+(v) + H_2$ to compare with our work. We can, however, compare our results with those of Vance and Bailey [32], who obtained cross sections for an unselected beam of H_2^+. If we assume their distribution of vibrational levels was Franck-Condon, we estimate an averaged cross section of 9.4 Å^2 at 8 eV. This compares favorably with their value of 10.9 Å^2 as well as with the value of 10.3 Å^2 measured by Lee et al. [36]. The increase of the CID cross sections with vibrational levels is not surprising, because the dissociation energy decreases rapidly with v. Nevertheless, it is impressive that CID cross sections as large as 7 Å^2 are obtained in this system. Simple theoretical treatments of CID, such as the optical model of Levine and Bernstein [38], predict that the cross sections should increase rapidly with translational energy. This is clearly not the case for $H_2^+ + H_2$.

The time of flight distribution for H^+ products produced from three different vibrational levels v is shown in Fig. 11. The time-of-flight is inversely proportional to the velocity component of the products along the initial relative velocity. For each value of v a different bimodal distribution is obtained. We attribute the presence of two peaks to electron exchange between the H_2^+ and H_2 prior to dissociation. Assuming the charge transfer mixing is complete, one expects two H^+

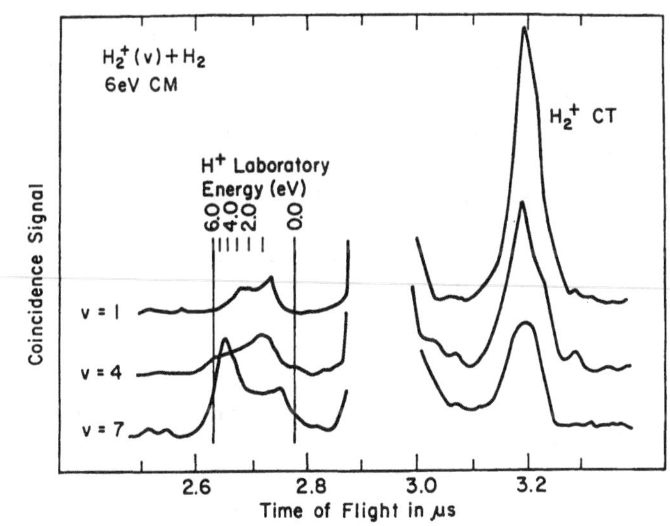

Fig. 11. The time-of-flight distributions of H^+ ions produced in collisions of $H_2^+(v) + H_2$ at E_{cm} = 6 eV for v = 1, 4, and 7 [15]. Also shown are the "slow" H_2^+ ions produced by charge transfer (CT). The large central peak of unreacted H_2^+ ions is omitted here. The TOF's of the H^+ ions have been converted to laboratory energies; the scale is shown in the Figure

peaks of equal intensity, a "fast" peak from dissociation of the projectile H_2^+, and a "slow" peak from dissociation of a target gas molecule.

Eaker and Muzyka [39] have performed a trajectory-surface hopping calculation on the $D_2^+ - H_2$ system, but restricted to the two lowest potential energy surfaces. They observed the electron transfer between the two molecules prior to dissociation, but their CID cross section for $D_2^+(v = 3) + H_2$ at 4 eV was only 0.9 $Å^2$, about three times smaller than our result. In addition, their calculated ratio for H^+ to D^+ products was 8:1, whereas our result was 1:1. A possible explanation of the discrepancy [15] is a mechanism in which the second excited state of the $[H_2H_2]^+$ system is excited during the collision. This state, which was not considered by Eaker, is repulsive for the H_2^+ moiety, and would give slow (fast) H^+ ions if electron transfer did (did not) occur during the first part of the collision. This suggestion remains to be tested by theory.

7 Conclusions and Future Prospects

The combination of the TPEPICO method with synchrotron radiation has provided an ideal way to prepare state-selected ions which can be used to study ion-molecule reactions. During the past four years we have used this method to study the role of vibrational energy in near-resonant charge transfer reactions, proton transfer reactions, and collision-induced dissociation processes in selected small systems. Many of the results have been unexpected and have stimulated several new theoretical efforts. As a result, we have a deeper understanding of both charge transfer and collision-induced dissociation. In addition, we have a better understanding of the competition betweeen chemical reaction and charge transfer in systems like $N_2^+ + H_2$.

The present apparatus allows us to study the range of collision energies between 2 and 100 eV. This range bridges the gap between high energy beam experiments and thermal drift-tube experiments. In contrast with those methods, however, the present data acquisition rate is quite small, with typical signals of 1 count/sec. This is an inherent limitation in the type of coincidence experiments we do. Until now the situation was even worse, because it was necessary to subtract a substantial false coincidence signal from the measured signal. The poor signal-to-noise ratio from such a procedure severely limits the number of experimental parameters such as collision energy and target gas which can be varied in a typical set of experiments. Thus, it is vital in the future to improve the ratio between true and false coincidences.

Recently, great progress has been made on this problem in our laboratory by M. Lavollée and G. Henri [40]. They have developed a pulsing method to eliminate more than 90% of the correlated ions, thereby increasing the signal-to-noise ratio by more than one order of magnitude. They have used this technique to study collisions of excited O^+ ions with N_2, a reaction of great atmospheric interest. These atomic ions are produced in small quantities by dissociative photoionization of O_2, but the experiments are straightforward thanks to the new pulsing method.

We plan to continue these experiments when the new synchrotron ring SUPERACO opens in Orsay. The new pulsing method will be used to reduce false coincidences.

An octopole ion beam guide is being constructed, which will permit us to study collision energies down to 0.1 eV. In addition, we have made new position sensitive detectors, which will allow us to measure complete energy and angle differential cross sections. We anticipate that future results will be even more exciting than those we have described in this paper.

Acknowledgements. The authors would like to thank all those who participated in the experiments and made this project so successful. Particular thanks go to T. Baer and T. R. Govers, who contributed to the original design; and to O. Dutuit, M. Lavollée, H. Fröhlich, G. Henri, and J. B. Ozenne for their many contributions to these experiments. We gratefully acknowledge support from the staffs of LURE, LCAM, and LPCR. Financial support was obtained from the "RCP Dynamique Reactionelle des Systems Simples." Finally, we are very grateful to the theoreticians at Orsay, especially P. Archirel, B. Levy, G. Parlant, and M. Sizun, and abroad, H. Baer, for their fruitful collaboration with us on this project.

Support from CNRS-NSF exchange program is gratefully acknowledged.

8 References

1. Baer T, Guyon PM, Nenner I, Tobché-Fouhailé A, Botter R, Ferreira LFA, Govers TR (1980) J. Chem. Phys. 72: 6587
2. (a) Govers TR, Guyon PM, Baer T, Cole K, Frohlich H, Lavollée M (1984) Chem. Phys. 87: 373; (b) Guyon PM, Govers TR, Baer T (1986) Z. Phys. D 4: 89
3. Kato T, Tanaka K, Koyano I (1982) J. Chem. Phys. 77: 834
4. (a) Liao CL, Xu R, Ng CY (1986) J. Chem. Phys. 85: 7136; (b) Shao JD, Li YG, Flesch GD, Ng CY (1987) J. Chem. Phys. 86: 170
5. (a) Spalburg MR, Los J, Gislason EA (1985) Chem. Phys. 94: 327; (b) Spalburg MR, Gislason EA (1985) Chem. Phys. 94: 339
6. Parlant G, Gislason EA (1986) Chem. Phys. 101: 227
7. Archirel P, Levy B (1986) Chem. Phys. 106: 51
8. Nikitin EE, Ovchinnikova MY, Shalashilin DV (1987) Chem. Phys. 111: 313
9. Henri G, Lavollée M, Dutuit O, Ozenne JB, Guyon PM, Gislason EA (1988) J. Chem. Phys. 88: 6381
10. (a) Gislason EA, Henri G, Labollée M, Dutuit O, Ozenne JB, Guyon PM (1988) Disc. Faraday Soc. 84: (1987); (b) Henri G, Lavollée M, Dutuit O, Ozenne JB, Gislason EA, Guyon PM (1988) In: Geddes J (ed) XV ICPEAC Book of Abstracts, Elsevier, Amsterdam p 674
11. Govers TR, Guyon PM (1987) Chem. Phys. 113: 425
12. Gislason EA, Guyon PM (1987) J. Chem. Phys. 86: 677
13. Sizun M, Parlant G, Gislason EA (1988) J. Chem. Phys. 88: 4294
14. Cole SK, Baer T, Guyon PM, Govers TR (1984) Chem. Phys. Lett. 109: 285
15. Guyon PM, Baer T, Cole SK, Govers TR (1988) Chem. Phys. 119: 145
16. Liao CL, Ng CY (1986) J. Chem. Phys. 84: 197
17. Cole SK, DePristo AE (1986) J. Chem. Phys. 85: 1389
18. Schatz GC, Badenhoop JK, Eaker CW (1987) Int. J. Quant. Chem. 31: 57
19. Gislason EA, Parlant G (1987) Comments At. Mol. Phys. 19: 157
20. Ng CY (private communication)
21. Tanaka K, Durup J, Kato T, Koyano I (1981) J. Chem. Phys. 74: 5561
22. North GR, Leventhal JJ (1969) J. Chem. Phys. 51: 4236
23. Hierl PM, Stratton LW, Wyatt JR (1972) Int. J. Mass Spect. Ion Proc. 10: 385
24. (a) Chupka WA, Russell ME (1968) J. Chem. Phys. 49: 5426; (b) D'Amico PM (1968) ACM Student Report
25. Turner T, Dutuit O, Lee YT (1984) J. Chem. Phys. 81: 3475

26. Whitton WN, Kuntz PJ (1976) J. Chem. Phys. 64: 3624
27. Joseph T, Sathyamurthy N (1984) J. Chem. Phys. 80: 5332
28. Connally CM, Gislason EA (1972) Chem. Phys. Lett. 14: 103
29. Schneider F, Havermann U, Zulicke L, Herman Z (1977) Chem. Phys. Lett. 48: 439
30. Stine JR, Muckerman JT (1978) J. Chem. Phys. 68: 185
31. Menendez MG, Thomas BS, Bailey TL (1965) J. Chem. Phys. 42: 802
32. Vance DW, Bailey TL (1966) J. Chem. Phys. 44: 486
33. Champion RL, Doverspike LD, Bailey TL (1966) J. Chem. Phys. 45: 437
34. Anderson SL, Houle FA, Gerlich D, Lee YT (1981) J. Chem. Phys. 75: 2153
35. Campbell FM, Browning T, Latimer CJ (1981) J. Phys. B 14: 3493
36. Lee CY, DePristo AE, Liao CL, Liao CX, Ng CY (1985) Chem. Phys. Lett. 116: 534
37. (a) Lee CY, DePristo AE (1983) J. Am. Chem. Soc. 105: 6775; (b) (1985) J. Chem. Phys. 80: 1116
38. Levine RD, Bernstein RB (1971) Chem. Phys. Lett. 11: 552
39. Eaker CW, Muzyka JL (1985) Chem. Phys. Lett. 119: 169
40. Lavolle M, Henri G (1988) In: Geddes J (ed) XV ICPEAC Book of Abstracts, Elsevier, Amsterdam, p. 701

Time-Resolved X-Ray Absorption Spectroscopy Using an Energy Dispersive Optics: Strengths and Limitations

Alouin Fontaine, Elisabeth Dartyge, Jean Paul Itie*, Alain Jucha, Alain Polian*, Helio Tolentino, Gerard Tourillon

LURE (Lab. CNRS, CEA, MEN) Bât. 209D, 91405 ORSAY Cedex, FRANCE
Physique des Milieux Condenses (C.N.R.S. — U.A. 782)

Table of Contents

* Université P & M Curie, T13 E4, 4 place Jussieu, F-75252 Paris Cedex 05

X-ray Absorption Spectroscopy has undergone a great theoretical and experimental development in the last years. This technique has proved to be a powerful tool in elucidating huge number of questions in materials science. Great interest exists in time-resolved experiments achieved with extreme energy resolution and energy scale stability taking full advantage of the strong correlation between the stereochemical environment of the absorbing atom and the exact shape and position of the absorption edge.

Fast energy dispersive X-ray spectroscopy allows in situ observations with data collected in a short time. Thus structural modifications are easily found and moreover, this scheme provides high energy resolution. However, quantitative analysis for very dilute systems are photon limited. A great benefit is expected from the forthcoming storage ring (ESRF) which should be able to give flux by at least 3 orders of magnitude greater. Only a few detection systems have been to compete with this creditable performance by synchrotron radiation source.

Nowadays the main limitation concerns very low-concentration samples since it is no longer possible to use the dispersive geometry. The fluorescent detection has proven to be efficient in photosynthesis study at concentrations of about 100 μmol but the use of decay channels is irrelevant for the dispersive schee since they are not dependent on the energy of the photon which creates the photoelectric core hole.

1 Introduction

Using the combination of X-ray energy dispersive optics and a position-sensitive detector able to work under high flux conditions we were able to proceed with both *in situ* and *time dependent* investigations. The cooled photodiode array is the unique tool to gives a good spatial resolution and large dynamics [1]. Several papers have evidenced already what are the major benefits one can expect for chemistry [2], physics [3], material science [4], and biophysics [5, 6, 7]. The goal of the present paper is to discuss how valuable the energy dispersive scheme is for science. In addition, we will try to give a tentative prospect of the advantages expected from the forthcoming European source.

But we would like to apologize for the incompleteness of this report which is not an extensive review, but rather a selection of typical experiments carried out in our laboratory using the dispersive scheme. In particular, we do not report the important experiments, including stopped-flow measurements developed by the Japanese scientists (lead by T. MATSUSHITA) who have been the very first to combine synchrotron radiation and dispersive optics to perform X-ray absorption spectroscopy. The readers interested can easily overcome this deficiency by looking through the proceedings of the three last EXAFS conferences held at Stanford (1984) Fontevraud (1986), and Seattle (1988) as mentioned in the reference list.

The impact on science of this new tool is illustrated by three experiments. The first one deals with an in situ time-resolved electrochemical reaction. The second illustration draws a great benefit from the energy scale stability and the extreme sensitivity of the detection scheme for the in situ investigation of oxygen uptake and removal of oxygen in high T_C superconductor. The last example points out the definitive advantage of the smallness of the beam size at the sample position for high pressure experiments. This is a major piece of new sience in our spectroscopy thanks to the recent optical improvements we have implemented in our station.

2 Energy Dispersive Scheme

Only 4 ms are needed to collect a full EXAFS spectrum spread over 500 eV when a triangular shaped Si bent crystal is used as a focusing dispersive optics and a photodiode array as a position sensitive detector. Each of the 1024 sensing elements of the photodiode array transforms an average of 10^5 8 keV-X-ray photons into 8.8×10^7 electron-hole pairs. In total, each frame is made of 10^8 photons which can be repeated at 550 Hz or 220 Hz according to used analog-digital converter (10 or 12 bits). It is worthwhile summarize the advantages and limitations related to this scheme:

1) The copper XANES spectra shown of Fig. 1 have been recorded using a Si (111) crystal to give an account of the energy resolution which is as good as it is with a step-by-step scan using a double detuned crystal optics. For a large X-ray source, the detector has to be set at twice the crystal-to-focus distance to achieve the optimized energy resolution [8]. For small sources such as those of ESRF this restriction does not exist, provided the detector is not close to the focus point.

2) In addition, because of the lack of mechanical movements, once the optics is tuned for a given absorption edge, the energy scale is very stable. This enables extreme

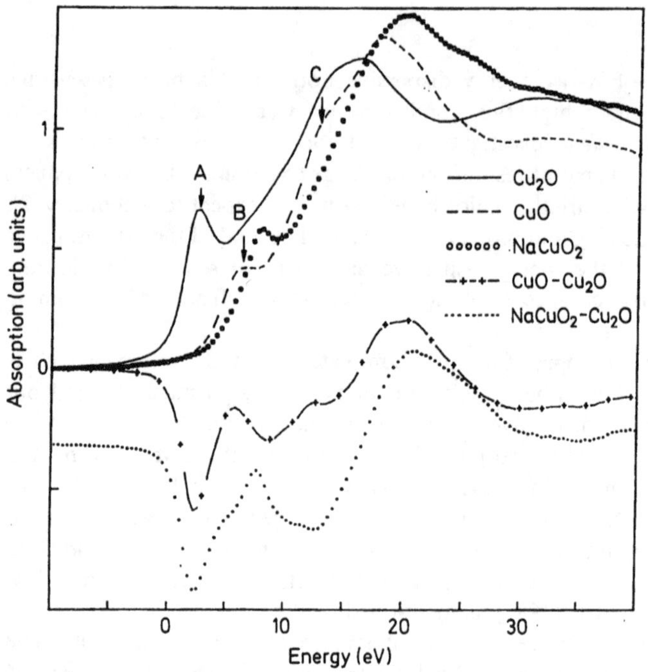

Fig. 1. Copper K-edge XAS spectra of Cu_2O, CuO and $NaCuO_2$ whose formal oxidation states are Cu(I), Cu(II) and Cu(III). The differences between the monovalent spectrum and the others are plotted, showing that the A peak ($1s^2\ 3d^{10}\ 4p^0 \rightarrow 1s^1\ 3d^{10}\ 4p^1$) is very well separated from the other features in the spectra

sensitivity meaning that even very minute energy shifts of the absorption threshold induced by chemical change can be detected accurately.

Most of the current optics using Synchrotron Radiation diffracts in the vertical plane and thus is sensitive to vertical bouncing of the beam. The horizontal optical plane of the dispersive scheme combines this extra advantage which helps to keep superior energy resolution since the orbit seems to show a better stability in the horizontal direction. Owing to the horizontal polarization of S.R., one must consider the $|\cos(2\theta)|$ attenuation factor which reduces the Darwin width of the crystal. This results in a lower reflectivity and an improved energy resolution as well.

3) The linearity and the dynamics of the photodiode array give very good signal/noise ratio. The signal is essentially statistically limited as proved by Fig. 2 which shows six spectra of copper collected in 1, 3, 8, 32, 320, and 1920×6.8 ms, respectively using Si (111).

4) Since this detector is made of silicon, a disadvantage is the decrease of the quantum efficiency for high energy X-ray photons. An advantage is the relative unsensitivity for harmonics which facilitates work under high absorbance ($\mu t \sim 2 \rightarrow 6$). Nevertheless, our spectrometer includes a mirror put just behind the sample to reject the harmonics. It also works as horizontal slits of high definition.

5) Combined to time-resolved capability (60 spectra, each of them being made of one or more U up to 32 — frames can be collected in one single shot) the in-situ

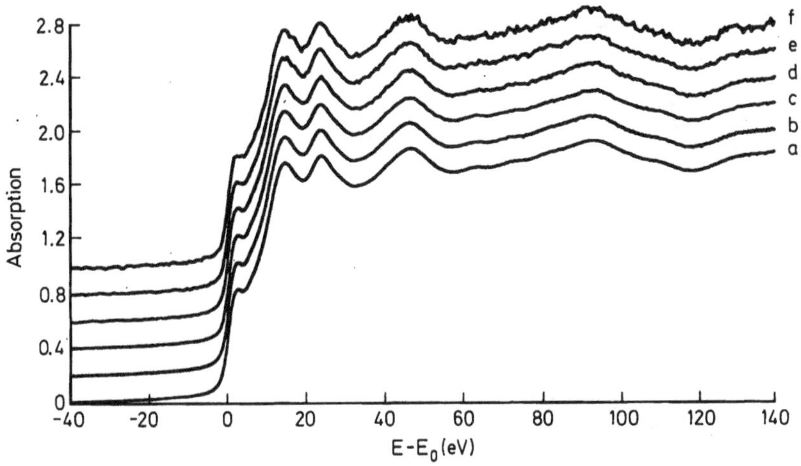

Fig. 2. EXAFS spectra for Copper metallic foil with a Si(111) bent crystal as a dispersive optics collected in f: 6.7 ms, e: 20.2 ms, d: 54 ms, c: 216 ms, b: 2.16 s, a: 12.98 seconds

observation permits a systematic investigation of the time-dependent system and an a posteriori decision about stages of interest.

6) One has to stress that the dispersive optics does not reflect more photons than a flat crystal. The main feature which differentiates a monochromator from the dispersive optics is that the local rocking curve of the bent crystal — almost constant in width — scans across a broad energy range when one goes from one side of the crystal to the other along the horizontal path of the beam span. On the contrary, for the step-by-step spectrometer at each elementary ray within the path the crystal shines at the same glancing angle and then all the Bragg reflected beams come from a single rocking curve which remains in constant position on the energy scale for the whole beam span.

3 Optical Considerations

In our experimental set-up at LURE, a 23 cm long bent crystal is illuminated by the white synchrotron X-ray beam, yielding a focus point which is the image of the source. The incident angle varies continuously along the crystal, providing a continuous change of the energy of the Bragg-reflected photons. This gives rise to a large energy band-pass. The energy-direction correlation is transformed into an energy-position correlation on the position sensitive detector (Fig. 3). In this geometry, the source to crystal distance p is large compared to the radius of curvature R of the crystal. The polychromatic focus point is within the socalled Rowland circle whose diameter is equal to the radius of curvature R. The well known relationship of the curved optics is given by

$$\frac{1}{q} + \frac{1}{p} = \frac{2}{R \sin \theta}$$

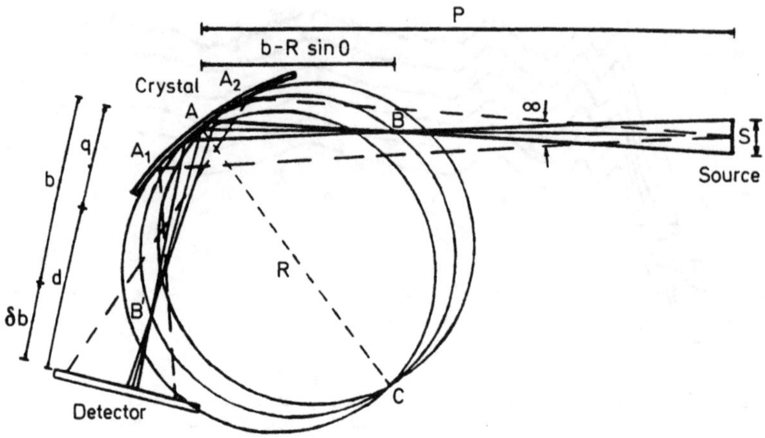

Fig. 3. Geometry of the curved crystal monochromator

and the obtained energy bandpass is

$$\Delta E = E \cot \theta \, \Delta\theta$$
$$= E \cot \theta (1/R - 1 \sin \theta/p)$$

where 1 is the iluminated length of the crystal.

Since there is no mechanical movement during data collection, the origins of possible shifts of the energy scale come from the radiation-produced thermal load on the crystal and from horizontal instabilities of the orbit of the positron bunch (averaged over a few thousands turns in the storage ring). The stability of the dispersive scheme has been measured at the copper K-edge crystal ($\sim$8980 eV) using a Si(111) monochromator and a reliability better than 30 meV was achieved after temperature stabilization [8]. This 30 meV can be converted into an angular fluctuation $\delta\theta = 2 \times 10^{-7}$ which, again is appreciable in terms of horizontal displacement of the source. This yields an upper magnitude of 3 μm, in the present situation where the source is at 15 m from the crystal.

Unfortunately, it often happens that the beam unstability causes energy shifts of the order of several hundreds when specific problems occur, such as, for example, unstabilties of the cooling water of the orbit correctors.

3.1 Energy Resolution

The energy resolution δE of an X-ray reflector comes from the derivative of the Bragg's law,

$$\delta E = E \cot \theta \, \delta\theta$$

where E is the nominal energy and $\delta\theta$ is the overall angular resolution. This includes:

i) the spatial resolution of the position sensitive detector P, which was measured to be almost two pixels, ,.e., P ≅ 50 μm. This contribution proceeds from the angle of view of the effective pixel from the polychromatic focus point, P/d;

ii) the intrinsec Darwin width of the rocking curve of the perfect crystal;

iii) the penetration depth of the incident X-ray into the crystal;

iv) the size of the X-ray source, whose contribution can be negligible under the conditions discussed below. As already pointed out, besides the polychromatic imaging at the focus point, which is the optimized position for the sample, there exists another type of focalization, called monochromatic focalization, which strongly controls the achievable energy resolution. The extended source provides, for each energy, a span of incident rays passing through the same point at the Rowland circle (Fig. 3). The rays passing through point B, for instance, diffract on the crystal in the Guinier conditions and focus at point B′, located at the distance b = R sin θ from the very part of the crystal which is reflecting these specified photons. This point is symmetric to B with respect to the diameter of the local Rowland circle. Therefore, if the detector is put at this monochromatic focus point, all the photons with this energy fall into a single pixel of the detector and the energy resolution is no longer limited by the extended source. If we consider the current situation in a synchrotron radiation beamline where the distance p is always much larger than q, one immediately finds that the optimized position b is almost twice the q value. Therefore, to optimize the energy resolution, the detector has to be installed at a distance which is almost two times larger than the crystal-to-sample distance.

The energy resolution as a function of the detector distance from a Si(311) monochromator is clearly seen by the attenuation of the strong white line at the arsenic K-edge of a chalcogenide sample As_2S_3 (Fig. 4). The position of the monochromatic

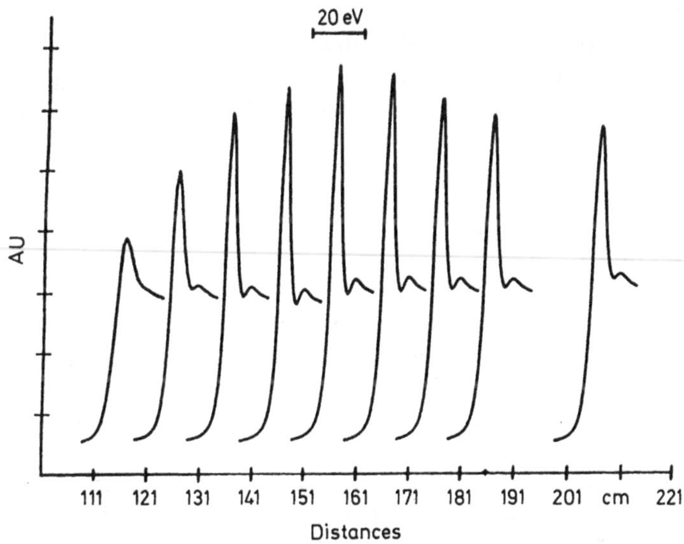

Fig. 4. Attenuation of the strong white line at the arsenic K-edge as a function of the detector position

focus is Bragg angle-dependent, not only because b varies with θ but also because this b value is measured from a point which runs along the crystal (points A, A1 and A2 in Fig. 3). If the detector is at the optimized position for a given energy, it is misplaced for the others. Fortunately, owing to both the limited source size (~6 mm) given by the magnetic dipole of the DCI (LURE) storage ring and the long source-to-crystal distance (~15 m), the field depth is large enough to keep a good resolution within a wide energy band-pass.

At the canonical CuK-edge, with Si(111) reflection and $q = 70$ cm, the field depth was verified to be approximately 12 cm. With the new low-emittance synchrotron radiation sources this last figure is no longer limited and the field depth is practically infinite.

Generally speaking, achieving a good resolution requires decreasing these three terms: E, $\cot\theta$ and $\delta\theta$. However, there is no choice for the energy, which is related to the binding energy of the core level of the probed atom. On the other hand, the Miller indices control both the Darwin width and, more importantly, the $\cot\theta$ value. Thus, real improvements in the energy resolution can only come from the optimization of the third term $\delta\theta$.

The extended source does not contribute if the detector is put at the monochromatic focus. It is the interplay between the other parameters that determines the resolution.

Owing to the good spatial resolution of our position sensitive detector (better than 50 μm), the dominant term is usually the Darwin width. Nevertheless, the Darwin width can be reduced by going to a higher order and/or asymmetric reflections and hence, the effect of the spatial resolution of the detector can become as important.

The penetration depth into the crystal broadens the width of the rocking curve but, owing to the small curvature used (always $R > 5$ m), it is controlled by the extinction and not by the absorption length. Thus, its contribution is negligible.

Another important parameter that may affect the resolution is the higher harmonic contribution from the Bragg reflector. A fused quartz mirror behind the monochromator has been currently used to reject this high harmonic contribution. Therefore, the energy resolution of the spectrometer is just limited by the Darwin width of the rocking curve and the spatial resolution of the position-sensitive detector [8].

3.2 Aberration-free Optics

The small size of the polychromatic image allows spatial resolved experiments. This advantage is exploited, for instance, when samples are very small [2] and when high pressure experiments are performed [9, 10]. However, when the required spatial resolution is smaller than the image delivered by the optics, the energy distribution within that image becomes crucial. The problem of optical aberration can be described as the nonhomogeneous distribution of the energy band-pass within the image (Fig. 5). The cylindrical optics, which is obtained by bending a triangular-shaped crystal, gives rise to an aberration proportional to the square of the length of the crystal ($\sim l^2$) [11]. This parabolic energy-position correlation is known as the U-like aberration. The ideal optics is elliptical, where the source and the image are located at the

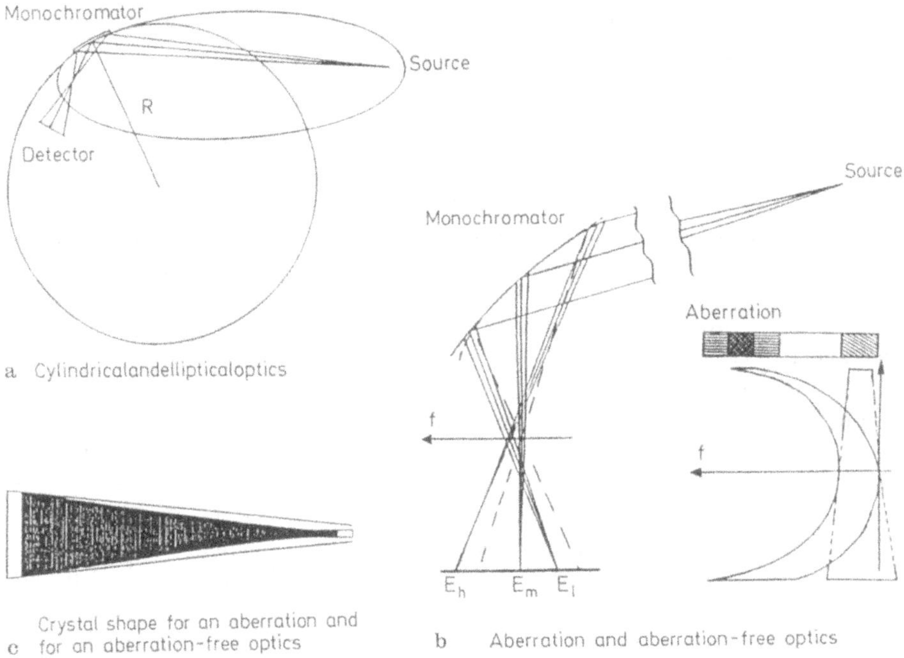

Fig. 5a–c. Main features of cylindrical and elliptical optics. The elliptical optics is the ideal one although the cylindrical gives rise to the U-like aberration.
a) cylindrical and elliptical optics;
b) aberration and aberration-free optics;
c) crystal shape for an aberration and for an aberration-free optics

focus of an ellipse and the bent crystal profile fits into an arc of this ellipse. This optics is aberration-free with the center of distribution of each energy being located at the same point (Fig. 5b). To tackle the aberrations of the cylindrically-bent crystal, corrections to the linear variation of the triangular shape were calculated and a set of Silicon crystals were tailored. The usual corrected shape of the crystal is displayed in Fig. 5c. The performance was investigated both by a fast ray tracing technique and experimentally. In order to measure the energy distribution within the image, a slit was swept through the polychromatic focus position (along the f axis in Fig. 5) and the intensities were recorded in the position-sensitive detector. It was verified that the size of the image diminished drastically to 400 μm, limited mostly by the demagnification factor ($\cong q/p \sim 0.7/14$) and the source size ($S \sim 6$ mm). For a Si(311) 24°-asymmetrically cut crystal optimized for the Fe K-edge, the extremely good energy distribution at the image enables us to record an iron K-edge through a 20 μm wide slit. It should be emphasized that with an aberration-free optics the spatial resolution achieved is no longer limited. By slitting down we just lose photons. This is a great improvement for investigation of powdered or non-uniform samples.

This result is also essential for high pressure experiments since the beam must go through a hole drilled in the gasket squeezed between the diamong anvils.

4 Time-Resolved in situ Observation of Electrochemical Inclusion of Metallic Clusters Within a Conducting Polymer

Conducting polymers represent an important class of materials in the fields of catalysis and energy stoarge, from both a fundamental and technological point of view. An appealing feature of polythiophene is its good stability against moisture and oxygen [13]. Furthermore, it is of very high purity since it is synthesized electrochemically without any calayst. Tourillon et al. [12] reported calalytic properties of poly-3-methylthiophene (PMeT) loaded with Cu and Pt aggregates. The size of the aggregates and the polymer-cluster interactions are obviously important to control the catalytic efficiency.

X-ray absorption spectroscopy is an attractive method of characterizing the growth and interaction of polymer-supported metal clusters since information on oxidation states, coordination geometry, and bonding angles can be determined [14]. The dispersive scheme offers the possibility of in situ investigations of growth mechanisms and kinetics of the electrochemical inclusion of copper into PMeT.

Several advantages combine very well for such an investigation:

1) the small size of the X-ray beam to probe the PMeT grafted on a Pt wire spatially.

2) the fast data acquisition to follow the time-dependent evolution of the copper clusters in the polymer.

4.1 XANES Approach to the Electrochemical Inclusion

In interpretation of XANES for any element beyond the first row of the periodic chart the 1*s* orbital, initial state of the K-edge absorption is essentially a hydrogenic wave function which is nearly a delta function centered at the origin. The energy shift between the K-edges of copper of different valencies reflects the difference between the final state configuration of these cations undergoing the X-ray absorption and is essentially due to the Coulomb interaction between the core hole c and the electrons in the d copper band (U_{cd}). In addition, holes in the d band enable the charge transfer between the absorbing atom and the ligand, the socalled shake down in the photoabsorption process. It exists for numerous Cu(II) compounds where the final state is a mixture of $|3d^9\rangle$ and $|3d^{10}L\rangle$ configurations as well-evidenced in CuO.

In addition, because of the dipole selection rule, the shape of the edge carries information concerning both the type and symmetry of the ligands. Hence, it has been proved that multiple scattering of the photoelectron by neighbors must be invoked to explain features close to the edges [14].

The grafting of poly-3 methylthiophene (PMeT) on the Pt wire involves the oxidation of the monomer, 3-methylthiophene, 0.5 M in CH_3CN + 0.5 M $N(C_4H_9)_4SO_3CF_3$ at 1.35 V/SCE (satured calomel electrode). The polymer is formed directly in its doped conducting state. This modified electrode is put in a 3 mm-thick electrochemical cell composed of a Teflon ring covered by two Kapton windows. An aqueous copper

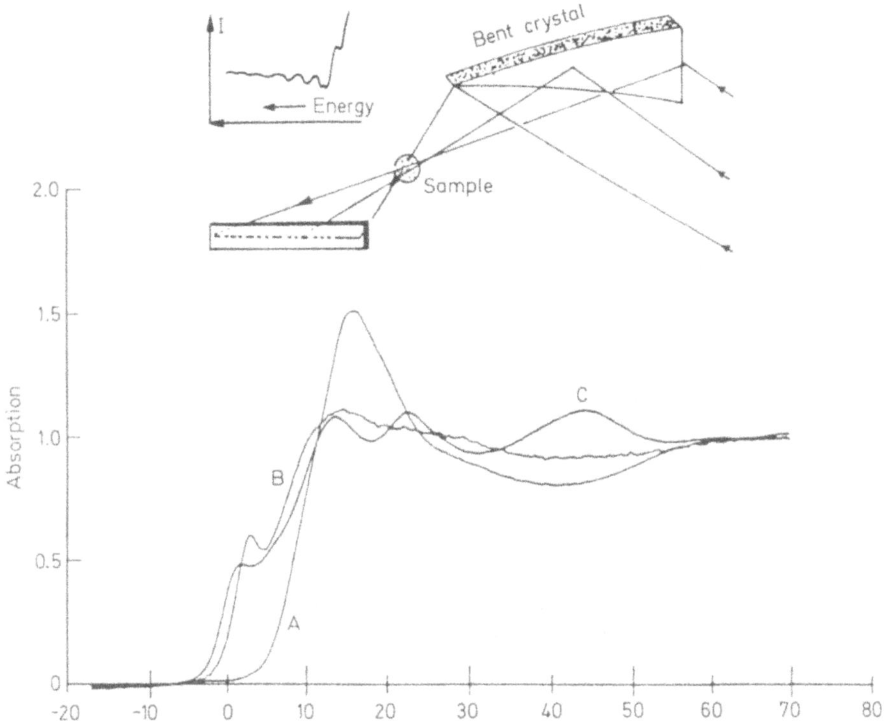

Fig. 6. CuK-edge absorption spectra of three copper species obtained on using a Si(311) crystal: —**A**, 50 mM aqueous $CuCl_2$ solution; —**B**, Cu^{1+} bipyridine complex in acetonitrile; **C**, copper foil (photon energy 8978 eV is taken to be the reference energy). Data has been collected in 9600 ms for curve **A**, 300 ms for curve **B**, and 460 ms for curve **C**

solution (H_2O + 50 mM $CuCl_2$, pH = 6) is then added to the cell. The grafted electrode and another Pt wire are used to develop the copper inclusions.

Figure 6 shows clearly the differences of the spectra which allow the estimation of the Cu^{2+}, Cu^{1+}, and metallic copper contents in this experiment. The evolution of the copper K-edge spectrum versus the cathodic polarization time was followed. The kinetics of the inclusion processes were determined from the time dependence of Cu^{2+}, Cu^{1+}, and Cu^0 concentrations.

Since the sample attenuates the X-ray beam by a factor of about 200, the collection time for each spectrum was chosen as being 3.6 s long data acquisitions were spaced at 7.2 s apart, twice the acquisition time of each spectrum. Figure 7 shows a series of spectra of the near-edge region of the Cu K-edge. A rapid shift of the copper edge towards lower energy (by 8.5 eV) appears, in addition to a small bump in the rise of the absorption just at the edge. This clearly shows that the Cu^{2+} ions transform into Cu^{1+} ions for the first step of the reaction. Since Cu^{1+} ions are unstable in aqueous solution, they must be stabilized by the polymer.

If the cathodic polarization is continuously applied, the bump in the rise of the absorption is shifted slightly from a value consistent with Cu^{1+} to the Cu^0 value. Also, the first two oscillations of the metallic spectrum which are separated by 9 eV

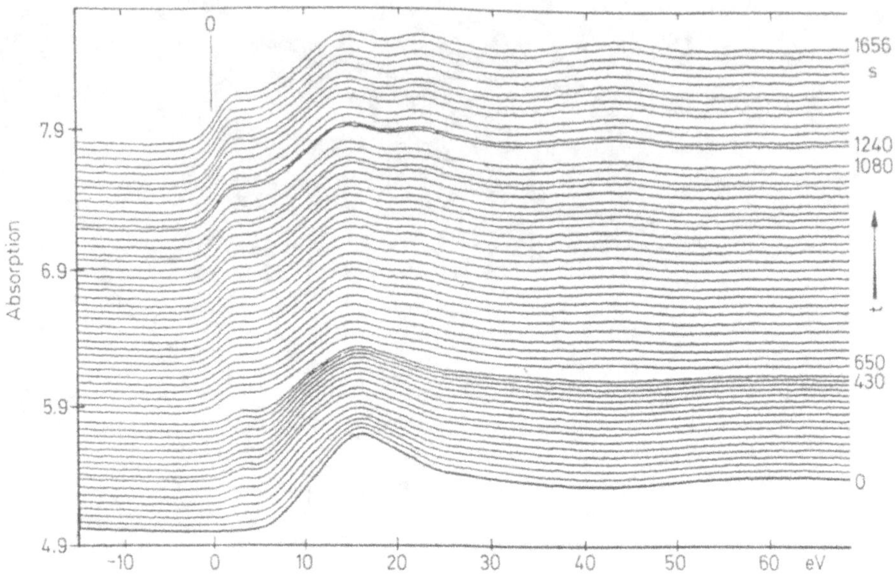

Fig. 7. In situ measurements of the evolution of the CuK-edge when pMeT is cathodically polarized in an $H_2O—CuCl_2$ 50 mM electrolytic medium

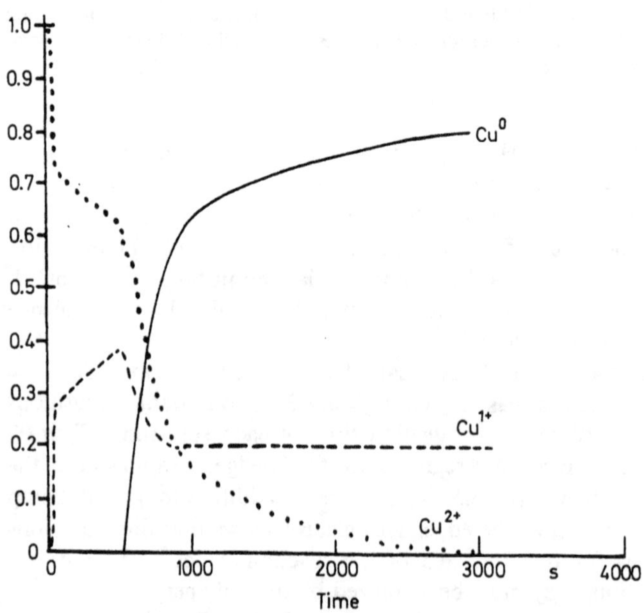

Fig. 8. Variation of the concentrations of Cu^{2+}, Cu^+, Cu^0 inside the polymer versus the polarization time

emerge fifteen minutes after starting the polarization (Fig. 7). This inclusion process is fully reversible. Reversing the potential from —4 V to 1 V leads to the metallic cluster dissolution into Cu^{1+} ions, followed by reappearance of Cu^{2+} ions on a similar time scale.

The time-dependent concentration of species (Fig. 8) can be determined by a simple analysis in terms of a linear combination of the XANES spectra of Cu^{2+}, Cu^{1+}, and metallic copper (Fig. 6). Three different kinetic domains must be considered:

1) The first one is a fast $Cu^{2+} \rightarrow Cu^{1+}$ transformation, where Cu^{1+} ions are stabilized by the polymer backbone and form complexes with the $SO_3CF_3^-$-ions. This assumption is consistent with our EXAFS data, which show that Cu^{1+} ions are surrounded by oxygen first which can come only from $SO_3CF_3^-$. The time constant of this rapid fixation is 27 s.

2) The second kinetic domain is characterized by a longer time constant [600 s (Fig. 8)], where the Cu^{1+} ion concentration increases form 25% to 40%. Since the doping level of PMeT is equal to 25%, no more SO_3CF_3-ions are available for this process. Thus, the newly synthesized Cu^{1+} ions should come from a direct interaction with the polymer backbone.

3) The last step is dominated by the metallic-copper cluster formation. We assume that in the initial process, all the accessible sulfur sites of PMeT are saturated. Then, in the absence of a stabilizing agent in aqueous solution, the monovalent copper ions undergo disproportionation to produce Cu^{2+} ions and metallic copper. Additional Cu^{2+} is then drained from the solution, resulting in an increase in the absolute copper content.

4.2 A Full EXAFS Investigation

The EXAFS data have been collected using a Si(111) crystal tuned at the copper K-edge with a wide energy band pass [15] ($\sim$500 eV.). A selected set of the k^3-weighted Fourier transforms using a 350 eV-wide window are shown in Fig. 9. One can observe as a first evolution (Fig. 9a to b) a shift of the principal peak towards the short distances, as well as a broadening ($\Delta t = 80$ s). After 168 s this peak is split into two components (Fig. 9c), whose intensity ratio inverts with the polarization time. The fast evolution of the system is illustrated in Fig. 10 which corresponds to two consecutive spectra separated by only 8 s. Later a shoulder appears at long distance on the main peak which becomes more and more pronounced. After 280 s, a single peak is observed which shifts continuously towards longer distances.

The intensity of the FT peak increases showing that the atoms of the first shell either have a larger backscattering amplitude or are in increasing number. At the end of the process, the characteristic FT of metallic copper is obtained. Figure 11 (a, b, c, d) shows the filtered back-transformed spectra of the first shell. These curves exhibit a continuous decrease of the amplitude of the oscillations with the appearance of a beat node at about 250 eV (Fig. 11c) directly related to the splitting of the Fourier transform. This beat node evidences that two different atoms with a π difference in their phase shifts contribute to the EXAFS oscillations. A direct explanation involves the O and S atoms in the first shell. This is consistent with the EXAFS characteristics drawn from two samples used as standards: the octahedrally hydrated Cu^{2+} solution and Cu_2S which exhibit a π difference in their phase shifts.

Thus, the first chemical step corresponds to the formation of Cu^{1+} ions complexed by O atoms given by the $(SO_3CF_3)^-$ dopant. The second step is clearly identified as the fixation of new synthetized Cu^{1+} ions by the S atom of the backbone of the polymer. These conclusions account for:

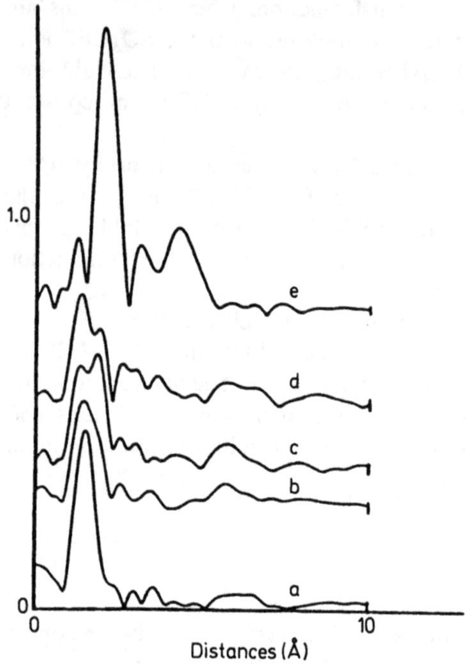

Fig. 9. Main steps of evolution of the k^3-Fourier transforms versus polarization time. a) 0 s, b: 100 s, c: 168 s, d: 200 s, c: 460 s

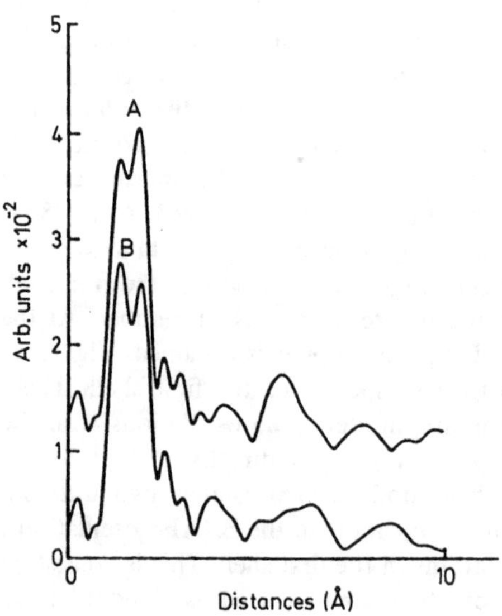

Fig. 10. k^3 Fourier transforms of two consecutive spectra separated by 8 sec. Note the inversion of the intensity ratio for the two peaks of the first shell. (A: 160 s, B: 168 s)

1) the broadening of the peak observed after ~80 s which results from the presence of Cu^{2+}-O, Cu^{1+}-O and Cu^{1+}-S bonds.

2) the formation of metallic copper clusters in the matrix which appears as the last step. The intensity ratio of the 3rd and 4th shell is inverted when compared to the bulk f.c.c. metallic copper: this implies that the early stage of metallic clustering involves a (111) platelet growth since the distances inside the (111) plane are observed preferentially to the distances between those planes.

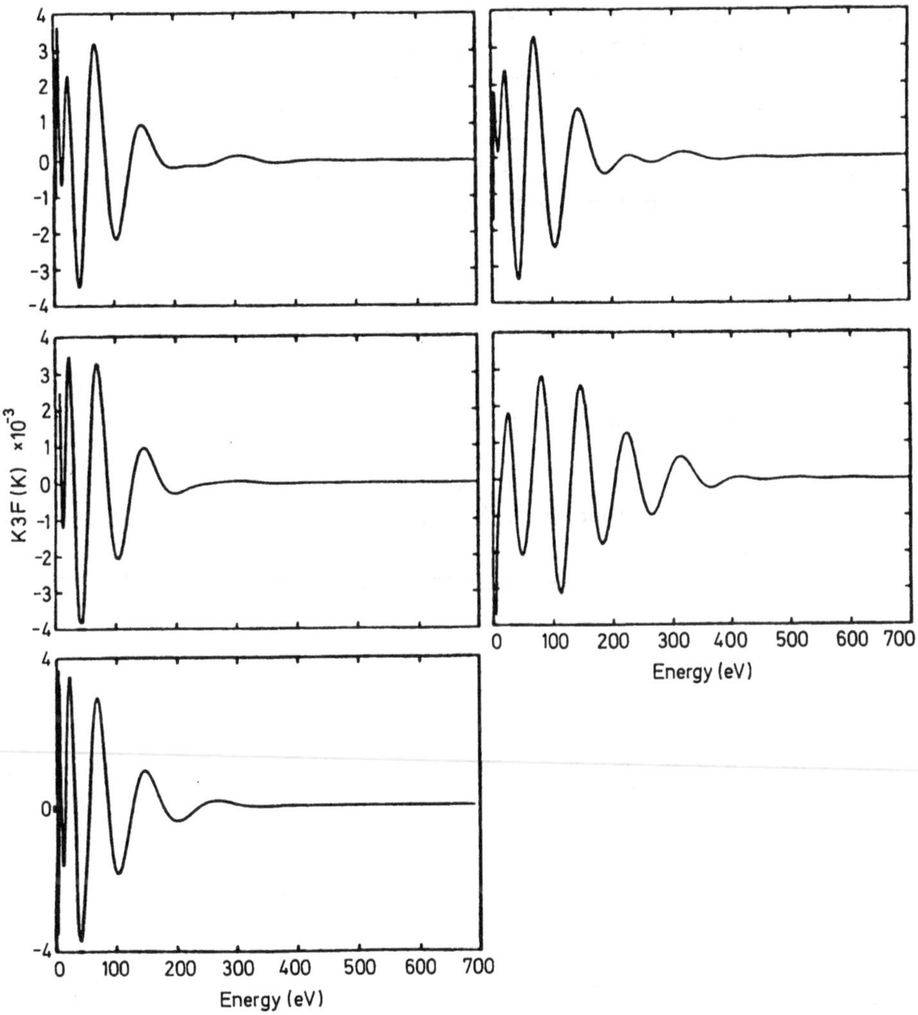

Fig. 11. Fourier backtransformed copper spectra. Steps of *Fig. 9*. Note that the splitted peak (c) results in a node of a beat at 220 eV, showing an exact compensation between the oxygen and sulfur contributions to the complex backscattering amplitude

5 New High T_C Superconductors

The $Y_1Ba_2Cu_3O_{7-\delta}$ family offers the unique possibility of varying continuously the electronic properties from a non-magnetic superconducting metal to an antiferromagnetic insulator, by merely changing the oxygen stoichiometry δ.

The knowledge of the electronic structure is a prerequisite for the explanation of the mechanism of superconductivity. Using X-ray Absorption Spectroscopy (XAS) in dispersive mode, we have studied the evolution of the Cu K-edge upon changing the in situ oxygen stoichiometry of $Y_1Ba_2Cu_3O_{7-\delta}$. Variations of less than 1% of the absorption spectrum, corresponding to different values of δ, could be reliably detected in the same sample. We observed directly the transformation of trivalent ($|3d^9\text{L}\rangle$ configuration) into divalent copper ($|3d^9\rangle$ configuration) upon increasing δ within the high oxygenated superconducting phase. When the compound becomes semiconducting a dramatic amount of monovalent copper, i.e., a $|3d^{10}\rangle$ configuration, is produced. Even for the transformation in the range of δ from 0.2 to 0.3 the ocurrence of a small fraction of this $|3d^{10}\rangle$ configuration is unambiguously observed, meaning that the hole concentration in that range is greater than that given by the chemical formula.

High energy spectroscopies, including XPS (mainly from Cu $2p$ and O $1s$ levels) [16–19], XAS at CuK and L_3 edges [19, 23], resonant photoemission [24] and EELS [25, 27] at 0 K edge, have been extensively applied in order to study the electronic ground state of these compounds. These former investigations have lead to the following clear-cut conclusions:

i) the electronic ground state is highly correlated,

ii) the trivalent copper exists mainly as a $|3d^9\text{L}\rangle$ configuration in the ground state ($|3d^9\text{L}\rangle$ meaning the occurrence of a hole in the copper *d* band and a hole in the oxygen *p* band), while no evidence for $|3d^8\rangle$ configuration (two holes in the copper *d* band) is found,

iii) empty states are unambiguously present in the oxygen *p* band in the metallic phase.

The analysis of our results on the $Y_1Ba_2Cu_3O_{7-\delta}$ compound, was enlightened by a comprehension of the main features of well characterized standard materials. Figure 1 shows the copper K edge spectra of Cu_2O, CuO and $NaCuO_2$ whose formal oxidation states are Cu(I), Cu(II) and Cu(III), respectively. Differences between them ($CuO-Cu_2O$ and $NaCuO_2-Cu_2O$) are also displayed. The structure A at 2.5 eV present in the Cu(I) spectrum reflects the dipolar allowed transition from the $1s$ state to the p_x and p_y degenerated empty states. According to a well-documented and systematic survey of XAS in many copper compounds [28], this A structure is always observed in linearly coordinated copper systems and its position comes from the $|3d^{10}\rangle$ configuration in the ground state. As can be seen, this structure is well separated from the others appearing in the Cu(II) and Cu(III) spectra, whose ground state has no such a $|3d^{10}\rangle$ configuration. The energy shift between these valencies reflects the difference between the final state configuration of these cations undergoing the X-ray absorption and is essentially due to the Coulomb interaction between the core hole *c* and the electrons in the *d* copper band (U_{cd}). In addition, holes in the *d* band enables the transfer of charge between the atom and the ligand, the so-called shake down in the photoabsorption process. This situation exists in the case of Cu(II) where the final state is a mixture of $|3d^9\rangle$ and $|3d^{10}\text{L}\rangle$ configurations.

The stoichiometry δ has been changed by controlling both the oxygen partial pressure and the temperature in a furnace able to keep a temperature stability of $\delta T \cong \pm 1$ °C from room temperature up to 1000 °C. The values of δ have been determined from the thermogravimetry measurements of Kishio et al. [29]. The data for 1 atmosphere oxygen pressure in our previous work [16] agrees with these measurements.

Here the emphasis is put on the difference of XAS spectra between two states corresponding to different values of δ. The stability of the energy scale which is better than 50 meV in this range of energy, yields a great sensitivity and reliability in the difference signals.

In the first step (path a in Fig. 12), a well loaded sample ($\delta \cong 0.05$) with a narrow superconducting transition at 92 K was progressively heated from room temperature (RT) up to 600 °C under 1 atmosphere of O_2 (i.e., up to $\delta \cong 0.20$). Note that in this range, the sample remains in the 90 K phase. The difference of XAS spectra (Fig. 13a) displays two maxima at about 1.5 and 10.2 eV, which increase progressively with the temperature. This difference compares very well with the derivative of the XAS spectrum from a $Y_1Ba_2Cu_3O_{6.95}$ sample (Fig. 14). This similarity can be easily explained if we assume that the absorption spectrum undergoes a slight shift δE towards the lower energy side. The most straightforward explanation is that, when δ increases, a small fraction of the ligand holes disappears, corresponding to a partial evolution from the $|3d^9\text{L}\rangle$ configuration into the $|3d^9\rangle$ one in the ground state. A rough simulation of this difference spectra reveals a shift $\delta E \cong 1.2$ eV; this shift is consistent with the difference of binding energy between these two configurations ($\delta E \cong 2$ eV) evaluated by Cu 2p XPS.

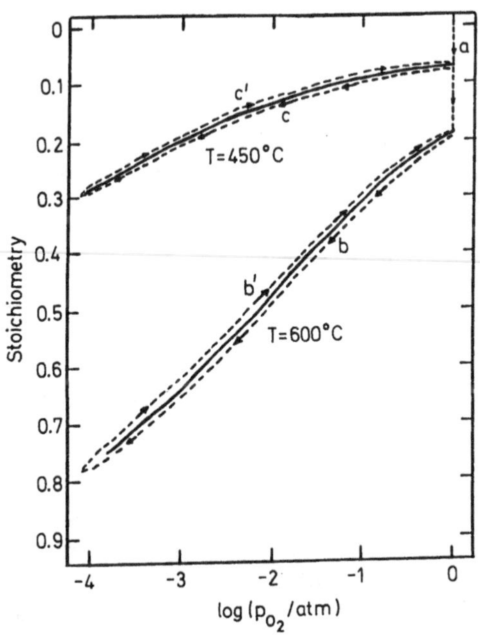

Fig. 12. Nonstoichiometry δ of $YBa_2Cu_3O_{7-\delta}$ sample as a function of the oxygen partial pressure for several temperatures. The paths (*dashed lines*) represent the conditions the sample was submitted during the in situ investigation. The values of δ are taken from reference [29]

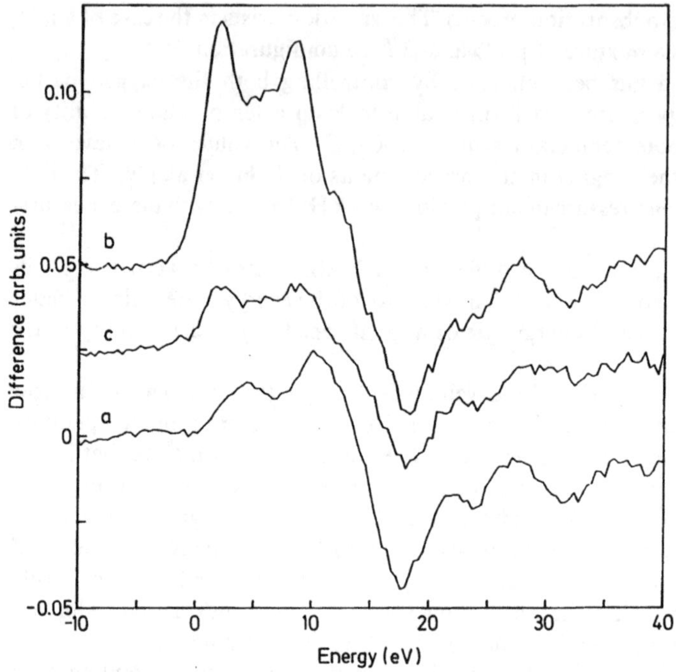

Fig. 13a–c. The difference between copper K-edge XAS spectra of $YBa_2Cu_3O_{7-\delta}$ annealed under the following conditions **a)** sample annealed from room temp. to 600 °C under 1 atm of O_2; **b)** sample annealed with oxygen partial pressure varying from 1 atm to 10^{-4} Torr for sample at 600 °C; **c)** id at 450 °C. The first step (a) yields essentially the transformation $|3d^9\text{L}\rangle \rightarrow |3d^9\rangle$ and the second one (b) essentially $|3d^9\rangle \rightarrow |3d^{10}\rangle$. The intermediate step (c) yields a mixture of both transformation. The A peak transition at 1.7 eV is clearly seen in the last two steps. An additional fact is that no variation in the quadrupolar pre-edge transition Q is observed

The transformation from the superconducting phase into the semiconducting one was achieved by changing the pressure from 1 to 10^{-4} atmosphere of oxygen at 600 °C. Thus, the stoichiometry δ was changed from $\delta \cong 0.2$ to $\delta \cong 0.8$ (paths b and b′ in Fig. 12). Now, the difference between XAS spectra evidences unambiguously the appearence of a strong additional feature at 1.7 eV (Fig. 13b), shifted by about —0.8 eV compared to the A peak of Cu_2O, which is interpreted as a signal coming from a $|3d^{10}\rangle$ configuration. This result shows clearly that the transformation of Cu(I) species becomes predominant in that range of δ. The removal of oxygen consists essentially in the transformation of the $|3d^9\rangle$, induced by square planar Cu(II), into the $|3d^{10}\rangle$ configuration. Nevertheless, it appears that partial transformation of the $|3d^9\text{L}\rangle$ into the $|3d^9\rangle$ configuration is still present, as indicated by the remaining features at about 4.5 and 10.2 eV.

After the 600 °C cycling the sample was cooled under oxygen atmosphere down to 450° and the same cycling of oxygen atmosphere was accomplished. Here, the oxygen deficiency varies from $\delta \cong 0.1$ to $\delta \cong 0.3$ (path c and c′ in Fig. 12). The difference between the XAS spectra (Fig. 13c) shows the additional feature at

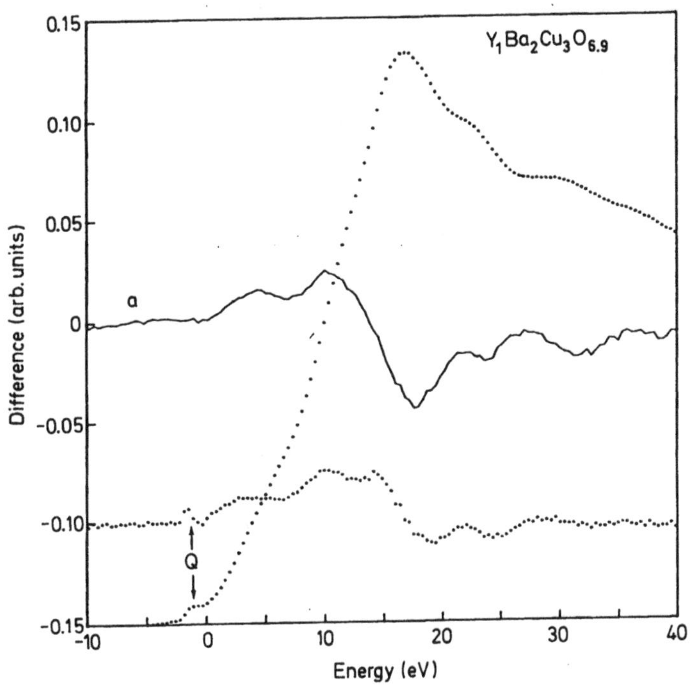

Fig. 14. Copper K-edge XAS spectrum of $YBa_2Cu_3O_{6.95}$ and its derivative (*dots*). The difference signal obtained by annealing under oxygen atmosphere is a derivative-like one and is due to a shift in the spectrum to the lower energy side

about 1.7 eV, even if a great contribution is due to the derivative-like signal of the first step. In fact, this proves that, in the considered range of δ, the main contribution comes from the transformation of the $|3d^9\text{L}\rangle$ into the $|3d^9\rangle$ configuration. However, there is some contribution from monovalent copper, i.e., $|3d^{10}\rangle$ configuration. This evidence of monovalent species is in apparent contradiction with the neutrality rule that predicts the appearence of Cu(I) only when δ is larger than 0.5. It should be noted that the reversibility of the oxygen stoichiometry variations is remarkably obtained, as illustrated in Fig. 15 where the differençes corresponding to the removal and uptake processes are displayed for both temperatures.

The presence of the $|3d^{10}\rangle$ configuration, which is most likely localized on Cu(I) sites, in the $Y_1Ba_2Cu_3O_{7-\delta}$ sample well beyond the tetragonal orthorhombic transition, implies that a) the hole concentration in the oxygen p band is higher than that given by the chemical formula and ii) that the tetra *ortho* transition should take place for δ larger than 0.5. Indeed, precise correlated magnetic and cristallographic measurements [30] have shown that this phase transition occurs at $\delta \cong 0.6$. Moreover, the initial oxygen uptake ($\delta > 0.75$), which drains electrons from the Cu(I) reservoir, does not change the Neel temperature. On the contrary, when the hole creation starts ($\delta < 0.75$) the antiferromagnetic order is strongly affected and the Neel temperature decreases, vanishing at $\delta \cong 0.4$.

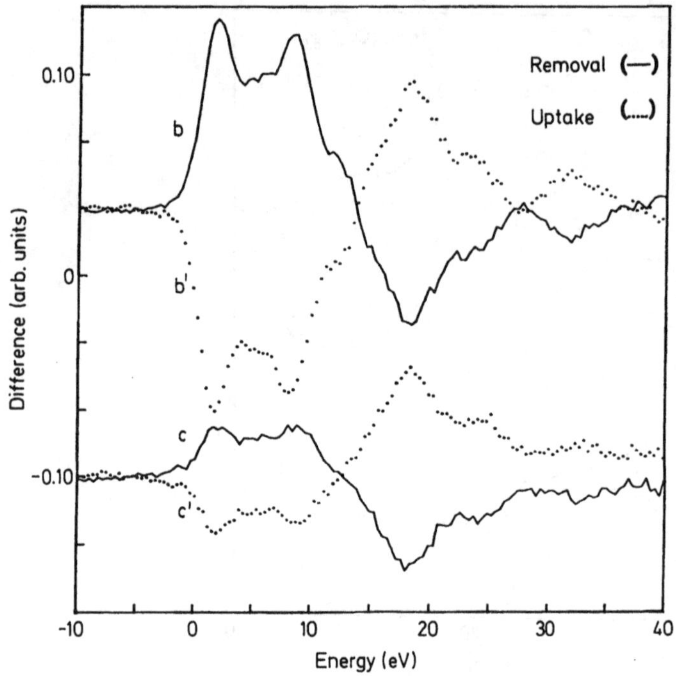

Fig. 15. Difference signals as obtained by cycling back and forth the oxygen pressure (1 atm → 10^{-4}) at 600 °C (b and b′) and 450 °C (c and c′). The uptake (. . .) and removal (———) of oxygen from sample is completely reversible, as evidenced by the perfect symmetry of the different spectra

6 High Pressure on Solid Krypton any GaP

Because of the spherical electronic shells of their constituent atoms, rare gas solids (R.G.S.) are the best candidates for the comparison of experimental results with theoretical predictions using various interaction potentials. XAS gives information about the interatomic distances and their mean square deviation which can be compared to data obtained by calculation or other experimental techniques. This section presents results on solid krypton under high pressure up to 20 GPa.

The high pressure cell was a classical Block-Piermarini diamond anvil cell, cryogenically loaded with Kr. The pressure was measured using the power 5 law ruby scale [31]. Because of the small size of the sample in the cell, 300 µm in diameter, only a part of the beam is usable.

In order to determine the nearest neighbor interatomic distance and the Debye-Waller factor, the classical EXAFS fitting procedure was used. Since at ambiant pressure and room temperature krypton is a gas, the backscattering amplitude and the phase shift have been obtained from the data at 15.7 GPa where the lattice parameter was known from X-ray diffraction [32]. Only the variation of the Debye-Waller factor can be measured.

The variation with pressure of the first neighbor interatomic distance obtained from the present EXAS measurement agrees with x-ray diffraction results. The thermal contribution to the Debye-Waller factor is determined by the mean square displacement of the central absorbing atom relative to its neighbors. Using self-consistent harmonic theory plus a cubic anharmonic term [33], Loubeyre calculated the mean displacement amplitude of the phonons in krypton [34] as function of the pressure, using Aziz's HFD-B pair potential [35]. Results of these calculations are shown in Fig. 16 (points). Since EXAFS data determine only changes in σ^2, a constant value of 0.0069 A^2 was added to experimental results (*stars* in Fig. 16). The comparison shows the remarkable agreement between both sets of data. The very fast variation

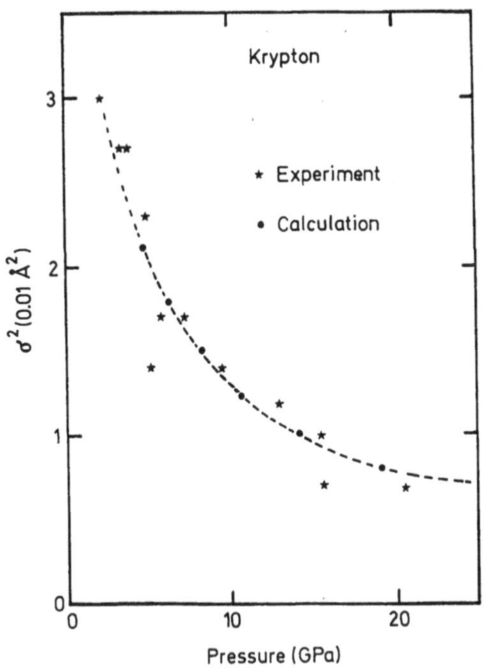

Fig. 16. Variation of the Debye-waller factor σ^2 with pressure

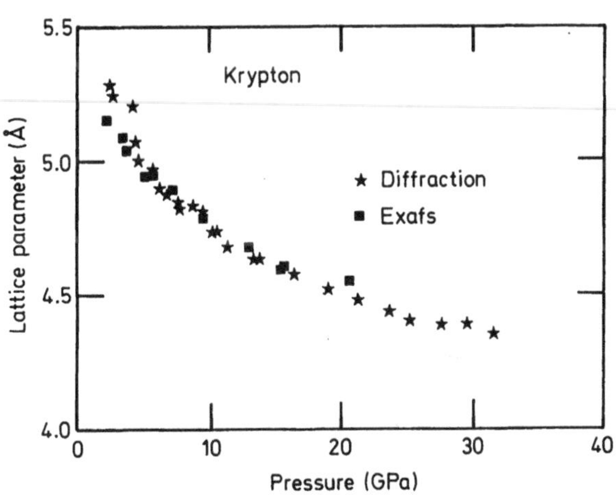

Fig. 17. The krypton lattice parameter as measured by EXAFS and energy dispersive diffraction

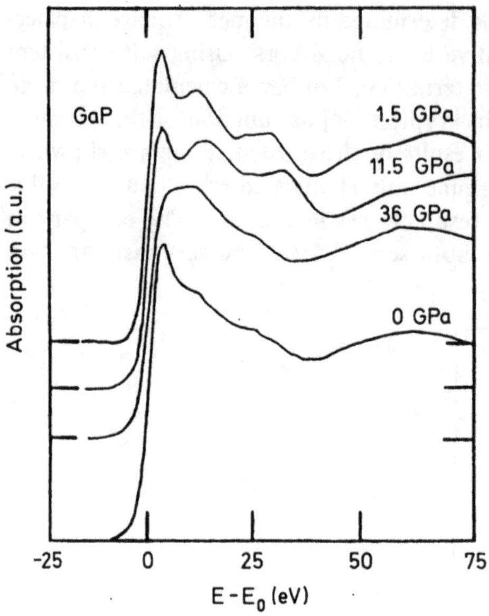

Fig. 18. Ga XANES spectrum measured at different pressures through a 250 μm-wide gasket hole

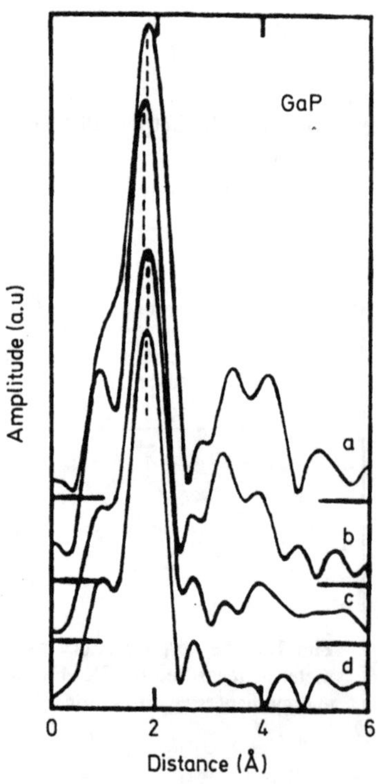

Fig. 19. Fourier transform of Ga EXAFS at different pressures: a) 1.2 GPa, b) 11.5 GPa, c) 36 GPa and d) 0 GPa after releasing pressure

of σ^2 with pressure at low pressure relates to the approach of melting and to the weak Van der Waals interatomic potentials.

A noticeable feature appears in the spectra, even at low pressure: The near edge part shows a double peak structure. The position in energy of the second peak is pressure-dependent and, within the experimental error, varies linearly with the lattice parameter (Fig. 17). Our set of data can provide useful information to deduce the density of the krypton bubbles formed after implantation in metals.

The phase transition of the semiconductor GaP under high pressure, occuring near 25 GPa yields another very clear illustration. These X-ray Absorption spectra were measured at the GaK edge, from O up to 36 GPa [9] (Fig. 18). We can observe the transition just by looking at the strong modification of the near edge region (XANES) of the spectra. The energy bandpass extended to about 400 eV without aberration enabled the analysis of the EXAFS spectra using the classical process of Fourier transform. Above 25 GPa, the first neighbor distance morease as expected. When pressure is released the sample become amorphous. This result contradicts in some way that which was claimed from diffraction.

7 Conclusion

The major interest of most of the studies is to show the new possibilities of in situ measurements since the dispersive scheme allows data to be collected at once. This is a major advantage over the step-by-step scheme.

For concentrated systems, in situ time-dependent experiments have been performed. The mechanisms and the kinetics of the electrochemical inclusion of copper particles inside an organic conducting polymer yield a complete illustration of the typical experiments one can consider. The time scale attainable of the order of a tenth of a second restricts this time-resolved spectroscopy to materials science where mass transportation is involved. ESRF can open the millisecond time scale giving access to more dilute samples and perhaps visualisation of changes of conformation of large molecules as found in biophysical related cases.

An additional advantage of the dispersive scheme comes from the focus of the beam which allows the use of very small apertures to carry out high pressure experiments [23]. Several other studies concerning catalysts, solid state amorphisation, biophysical systems or investigation of buried interfaces (adding the total reflection scheme to the energy dispersive mode) have been published in the proceedings of the International Conference held at Fontevraud (France) on July 1986, or Seattle (USA) on August 1988. Taking benefit of the whole stability of the spectrometer during data collection, we were able to collect spin-dependent X-ray absorption-signal which requires an accuracy of $\Delta\mu/\mu \sim 10^{-4}$.

8 References

Bianconi A, Inconnia L, Stipicich S (1982) (eds) Proceedings of the Int. Conf., Frascati, Italy, 13–17 September 1982, Springer Series in Chemical Physics Vol 27

Hodgson KO, Hedman B, Penner-Hahn JE (eds) Proceedings of the Int. Conf., Stanford, USA, 16–20 July 1984, Springer Series in Physics Vol 2

Petiau J, Lagarde P, Raoux D Proceedings of the IV Int. Conf. Fontevraud July 1986, Journal de Physique, C8, supp no 12, 42, Déc. 86

Stern E, Rehr J (eds) Proceedings of XAFS V Conference, Seattle August 1988, J. of Less Common Metals

1. Dartyge E, Depautex C, Dubuisson JM, Fontaine A, Jucha A, Leboucher P, Tourillon G (1986) NIM A246 p 452
2. Tourillon G, Dartyge E, Fontaine A, Jucha A (1986) PRL 57: 5, 506
3. Dartyge E, Fontaine A, Tourillon G, Cortes R, Jucha A (1986) Phys. Letters A, 113A: 7, p 384
4. Maire G, Garin F, Bernhardt P, Girard P, Schmitt JL, Dartyge E, Dexpert H, Fontaine A, Jucha A, Lagarde P (1986) Applied Catalysis 26: 305
5. Matsushita T, Oyanagi H, Saigo S, Kihara H, Kaminaga U In: Hodgson KO, Hedman B, Penner-Hahn JE (eds) EXAFS and Near Edge Structure III
6. Ascone I, Fontaine A, Bianconi A, Congui-Castellano A, Giovanelli A, Della Longa S, Momenteau (1987) In: Bianconi A, Congiu-Castellano A (eds) Springer series in Biophysics Vol 2 Biophysics and SR Synchrotron Radiation Ascone I, Bianconi A, Dartyge E, Della Longa S, Fontaine A, Momenteau M (1987) Biochimica & Biophisica Acta 915: 168.
7. Rohrer JS, Joo M-S, Dartyge E, Sayers DF, Fontaine A, Theil EC (1987) J. of Bio. Chem. 262: 13385
   Theil EC et al. Proc. 3rd Int. Conf. on Bioorg. Chem. Noordwijkerhout July 1987
8. Tolentino H, Dartyge E, Fontaine A, Tourillon G (1988) J. App. Crys 21: 15–21
9. Itie JP, Polian A, Jauberthie-Carillon C, Dartyge E, Fontaine A, Tolentino H, Tourillon G (1988) Proceedings of XAFS V Conference, Seattle
10. Polian A, Itie JP, Dartyge E, Fontaine A, Tourillon G (1989) Phys. Rev. B39, 3369
11. Ice GE, Sparks CJ (1984) Nuc. Inst. Med. 222: 121
12. Tourillon G, Dartyge E, Dexpert H, Fontaine A, Jucha A, Lagarde P, Sayers DE (1984) J. Electroanal. Chem. 178: 357
13. Tourillon G (1985) In: Skotheim T (ed) Handbook on conjugated electrically conducting polymers, (Marcelle Dekker, New York, vol 1, Chap. 9, p 294
14. Natoli CR (1983) In: Bianconi A, Inconnia L, Stipicich S (eds) EXAFS and near edge structures, Proceedings of the International Conference, Frascati, Italy, 13–17 September 1982, Springer Series in Chemical Physics, Springer Verlag, Berlin Heidelberg New York, vol 27 p 43
15. Dartyge E, Fontaine A, Tourillon G, Jucha A (1986)
    Proceedings of the International Conf. Fontevraud July 1986, Journal de Physique, 38, supp no 12, 42
16. Gourieux T, Krill G, Maurer M, Ravet M-F, Menna A, Tolentino H, Fontaine A (1988) Phys. Rev. B37: 751
17. Fuggle JC, Weijs PJW, Schoorl R, Sawatzky GA, Fink J, Nucker N, Durham PJ, Temmerman WM (1988) Phys. Rev. B37: 123
18. Fujimori A, Takayama-Muromachi E, Uschida Y, Gkai B (1987) Phys. Rev. B35: 6, 8814
19. Van Der Marel D, Van Elp J, Sawatzky GA, Heitmann D (1988) Phys. Rev. B37: 5136
20. Baudelet F, Collin G, Dartyge E, Fontaine A, Kappler JP, Krill G, Itie JP, Jegoudez J, Maurer M, Monod Ph, Revcolevschi A, Tolentino H, Tourillon G, Verdaguer M (1987) Z. für Physik B69: 141
21. Bianconi A, Congiu-Castellano A, De Santis M, Rudolf P, Lagarde P, Flank AM, Marcelli A (1987) Sol. St. Comm. 63: 11, 1009
22. Bianconi A, Budnick J, Flank AM, Fontaine A, Lagarde P, Marcelli A, Tolentino H, Chamberland B, Demazeau G, Michel C, Raveau B (1988) Physics Letters A127: 5, 285
23. Bianconi A, De Santis M, Di Cicco A, Flank AM, Fontaine A, Lagarde P, Katayama-Yoshida H, Kotani A, Marcelli A (1988) Phys. Rev. B38, 7196
24. Thiry P, Rossi G, Petroff Y, Revcolevskchi A, Jegoudez J (1988) Europhys. Lett. 5: 55

25. Batson PE, Chisholm MF (1988) Phys. Rev B37: 635
26. Nucker N, Romberg H, XiXX, Fink J, Gegenheimer B, Zhao ZX (1989) Phys. Rev. B39, 10, p 6619
27. Yuan J, Brown LM, Liang WY (1988) J. Phys. C21: 517
28. Penner-Hahn JE (1984) Thesis Stanford SSRL Report 84/03; Smith TA, Penner-Hahn JE, Berding MA, Doniach S, Hodgson KO (1985) J. Am. Chem. Soc. 107: 5945; Penner-Hahn JE, Scott RA, Hodgson KO, Doniach S, Desjardins SR Solomon EI (1982) Chem. Phys. Lett. 88: 595
29. Kishio K, Shimoyama J, Hasegawa T, Kitazawa K, Fueki K (1987) Jap. J. Appl. Phys. 26 L: 1228
30. Rossat-Mighod J, Burlet P, Jurgens MJ, Regnault LP, Henry JY, Ayache C, Forro L, Vettier C, Noel H, Potel M, Gougeon P, Levet JC (1989) Journal de Physique Int Conf on Magnetism 25–29 July 1988
31. Mao HK, Bell PM, Shaner JW, Steinberg DJ (1978) J. Appl. Phys. 49: 3276
32. Polian A, Besson JM, Grimsditch M, Grosshans WA (unpublished)
33. Loubeyre P, Levesque D, Weiss JJ (1986) Phys. Rev. B33: 318
34. Loubeyre P: private communication
35. Aziz RA, Slaman J (1986) Mol. Phys. 58: 679

Experimental Possibilities in Small Angle Scattering at the European Synchrotron Radiation Facility

Christian Riekel

European Synchrotron Radiation Facility B.P. 220, F-38043 Grenoble Cedex France

The European Synchrotron Radiation Facility is a dedicated 6 GeV synchrotron radiation source which will be operational by 1994. It will be optimized for a small beam emittance and the use of undulators and wigglers. A number of instruments proposed for small angle X-ray scattering will be reviewed.

1 Introduction

Small-angle X-ray scattering (SAXS) experiments using synchrotron radiation (SR) are performed at present mainly in the areas of real time scattering and anomalous dispersion.[1] Typical applications are the study of melting or recrystallisation of semi-crystalline polymers [4, 5], phase separation of alloys [6], muscle diffraction and stopped flow experiments on dissolved biopolymers [7, 8]. Anomalous dispersion has been exploited in order to determine partial structure factors in alloys [9] or polymers containing heavy atoms [10].

The European Synchrotron Radiation Facility (ESRF) in Grenoble (Fig. 1) — which is expected to come into operation in 1994 — will allow a considerable improvement of the experimental possibilities, especially in the extension of the accessible Q-range ($Q = 4\pi \sin \theta/\lambda$) and in the flux available at the sample. The aim of the present paper is to demonstrate this for three of the proposed SAXS-beamlines [11].

Fig. 1. Site and architectural plan for the ESRF in Grenoble. The plan shows the storage ring on the peninsula formed by the rivers Drac and Isère

[1] The term *small angle scattering* also comprises in the context of the present article low angle diffraction and low angle solution scattering. Reviews on instrumentation developed at existing SR-sources can be found in [1, 2, 3]

2 Matching of Source and Camera

2.1 Small Angle Resolution of a SAXS Camera

The discussion will be limited to the case of the focusing, pinhole SAXS-camera. The advantages of this type of camera are an optimized flux at the sample and the absence of slit smearing effects encountered in Kratky-cameras [12].

Figure 2 shows schematically the focussing geometry and the origin of diffuse scattering around the focus in the detector plane, due to scattering from the aperture slits (S_1) and the optical elements.

The largest observable d-spacing of such a camera depends on the limits of diffuse scattering in the detector plane. This limit — L — can be approximately calculated from the size of the aperture slits (S_1), the distance of the detector plane from the guard slits (L_2) and the size of the focus (a) [4]. In real space one can write:

$$L = \lambda/2[\sin(\tan^{-1} m/L_2)/2]^{-1} \tag{1}$$

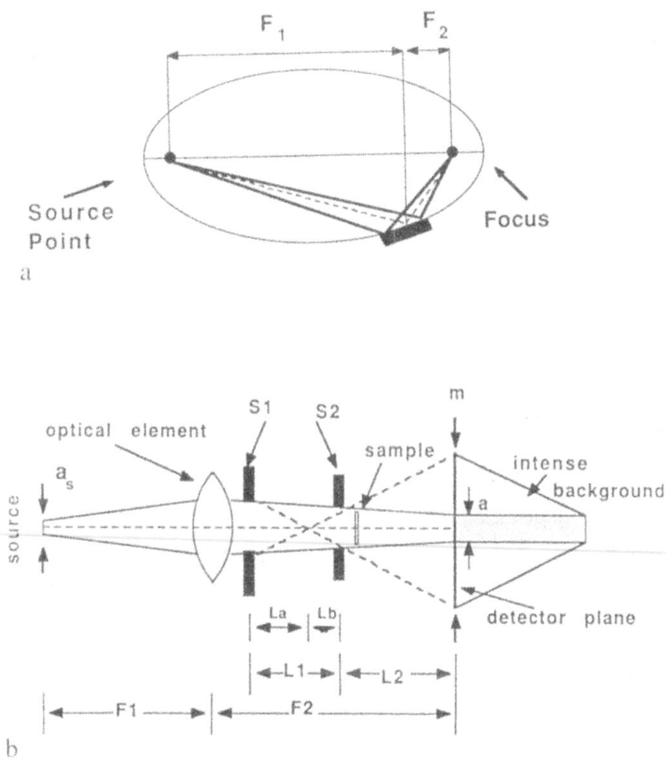

Fig. 2a. Focusing geometry of a SAXS-camera [11, 32]. **b.** Schematic design of a focussing SAXS-camera with a pinhole geometry. S_1 and S_2 are the aperture and guard slits. The optical element could be a mirror or a monochromator. m corresponds to the — idealized — extension of the diffuse scattering in the detector plane. a_s is the size of the source point and and a that of the focus. Note that the calculations of SAXS-resolution assume that $F_2 \approx L_1 + L_2$

where

$$m = (S_1/2L_a)(L_b + L_2) \quad (2)$$

$$L_a = (S_1/S_2) L_b \quad (3)$$

$$L_b = (L_1 \cdot S_1)/(S_1 + S_2) \quad (4)$$

$$S_2 = [(S_1 - a)/(L_1 + L_2)] L_2 + a \quad (5)$$

For the following discussion a coordinate system based on the source point will be used. (Fig. 3a) s is tangent to the electron beam at the source point. The electron (or positron) trajectory is in the x — s plane. P is an observation point defined by the angles θ and φ. The magnetic field vector is along z.

The size of the aperture slits in the horizontal (S_1^x) and vertical (S_1^z) directions is calculated from the photon beam source point size — x, z (σ) —, divergence — x′, z′(σ) and F_1, which is the distance source point to aperture slits:

$$S_1^x = 2 \cdot [x^2 + (x' \cdot F_1)^2]^{0.5}\ ; \qquad S_1^z = 2 \cdot [z^2 + (z' \cdot F_1)^2]^{0.5} \quad (6)$$

The size of the focus — a^x; a^z — is calculated from:

$$a^x = (F_2/F_1) \cdot a_s^x\ ; \qquad a^z = (F_2/F_1) \cdot a_s^z \quad (7)$$

where $a_s^x = 2 \cdot x$; $a_s^z = 2 \cdot z$.

The extent of the diffuse scattering in the detector plane is therefore dominated by the size of the aperture slits ($S_1^{x,z}$) and the length of the camera after the optical elements (F_2).

2.2 Requirements in Respect to the Source

2.2.1 Introduction

Realtime SAXS-experiments with perturbations which couple to the sample volume via an interface (e.g. heating, diffusion of chemical species) or mixing of components (e.g. stopped flow) demand an as small as possible sample volume in order to minimize the rise time of the perturbation. One is therefore interested in an as high as possible brillance of the source in order to keep $S_1^{x,z}$ as small as possible while utilizing the full photon beam. (Brilliance: Photons/(s · mm^2 · $mrad^2$ · BW); BW ≡ bandwith) This is particularly true for scattering on small samples like single biological fibrils. The requirements on the bandwidth are in the range $10^{-4} < \Delta E/E < 10^{-2}$ and depend on the sample [2].

For anomalous dispersion experiments, spectra have to be recorded with a bandwidth $\Delta E/E \approx 10^{-4}$ around a particular absorption edge of an atom. In order to cover a range of absorption edges, a wide spectral range of the source is necessary.

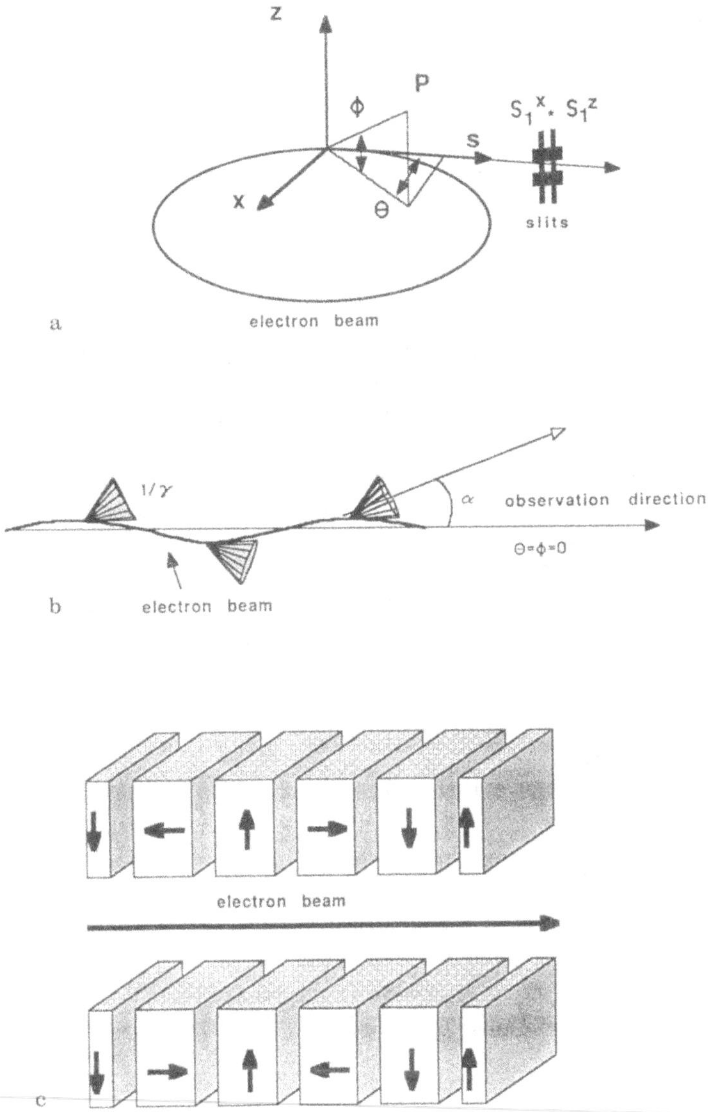

Fig. 3a. Coordinate system used to describe the motion of the electron beam and the emission of the photon beam. The magnetic field vector is along z. P is an observation point, the direction of which is defined by Φ and θ. S_1^x and S_1^z are horizontal and vertical slits. **b.** Schematic design of the electron trajectory and photon emission in a periodic magnetic structure. The photons are emitted into cones of opening angle $1/\gamma$. α is the angle between the cone axis and the observation direction $\theta = \Phi = 0$. **c.** Schematic design of one period — λ_u — of an insertion device, constructed in the permanent magnet technology. The arrows indicate the direction of the magnetic field [11]

2.2.2 Optimization of Brilliance

The brilliance of a photon-source can be optimized by a reduction of the emittance of the electron beam and the use of periodic magnetic structures (insertion devices) as radiation sources.

The horizontal — ε_x — and vertical — ε_z — emittances of the electron beam are related to spatial (σ_x, σ_z) and angular (σ_x', σ_z') deviations from the ideal orbit of the electrons via: [13]

$$\varepsilon_x = \sigma_x \sigma_x' ; \qquad \varepsilon_z = \sigma_z \sigma_z' \tag{8}$$

The photon beam parameters depend directly on the electron beam parameters. A reduced electron beam emittance will therefore also reduce S_1^x and S_1^z (Eq. 6). Size and divergence of the photon beam are, however, limited by diffraction effects when the beam emittance is reduced to values of about the photon wavelength $\varepsilon \approx \lambda$ [14].

Table 1 compares electron beam emittances for a number of existing and proposed SR-sources. One sees that the new generation of dedicated sources already closely approaches the theoretical limit.

Insertion devices are placed into straight sections of the storage ring. As shown schematically in Fig. 3b, the emitted photons add up in the observation direction. Note that the photons are emitted into cones of opening angle $1/\gamma$ (FWHH), where:

$$\gamma = 1957 \cdot E \tag{9}$$

E[GeV] is the energy of the ring.

Such devices consists of N periods of length λ_u. They can be characterized by the so-called deflection parameter K:

$$K = \alpha \cdot \gamma = 0.0934 \cdot B_o \cdot \lambda_u \tag{10}$$

Table 1. Comparison of electron beam emittances for selected existing and proposed SR-sources. For the proposed sources a 10% coupling between the emittances in the horizontal and vertical planes is assumed

Source	E (GeV)	ε_x (m · rad)	ε_z (m · rad)	status	Ref.
BESSY	0.8	$4 \cdot 10^{-8}$	$4 \cdot 10^{-10}$	operational	[13]
DARESBURY	2.0	$1 \cdot 10^{-7}$	$2 \cdot 10^{-9}$	operational	[16]
NSLS	2.5	$8 \cdot 10^{-8}$	$8 \cdot 10^{-10}$	operational	[13]
DORIS	3.7	$3 \cdot 10^{-7}$	$1.5 \cdot 10^{-8}$	operational	[15]
PEP	8.0	$8 \cdot 10^{-9}$	$8 \cdot 10^{-10}$	limited use	[18]
TRIESTE	2.0	$< 10^{-8}$	$< 10^{-9}$	under construction	[17]
PEP	4.0	$0.5 \cdot 10^{-10}$	$0.5 \cdot 10^{-11}$	proposed	[18]
ESRF	6.0	$6 \cdot 10^{-9}$	$6 \cdot 10^{-10}$	under construction	[11]
Diffraction limit		$\approx 10^{-10}$ ($\lambda = 1$ Å)			[14]

where α is defined in Fig. 3b. B_0[T] is the magnetic field. (λ_u in mm) For sufficiently small K-values (K $\lesssim$ 4), the emitted spectrum is no longer continuous but shows discrete peaks due to an interference between the emitted photons. Such an insertion device is called an undulator. For larger K-values, the spectrum resembles that emitted from a bending magnet, increased in flux, however, by the number of poles. Such an insertion device is called a wiggler. A schematic design of an insertion device-using permanent magnets- is shown in Fig. 3c.

Figure 4 shows the physical dimensions of the ESRF lattice. The circumference of the ring (844.39 m) is determined by the scaling of the beam emittance:

$$\varepsilon \approx E^2\Omega^2 \tag{11}$$

where Ω is the bending angle per bend magnet [14].

The energy of the ring is fixed to 6 GeV in view of extending the spectral range of the undulator fundamental up to 14.4 keV [11], which implies a large number of cells in order to keep the emittance low. In the case of the ESRF, the lattice contains 32 straight sections of which 29 can be used for insertion devices. The useful length of each straight section is 6.34 m. The first optical element can be installed just outside the shielding at $\approx$ 26 m from the source point.

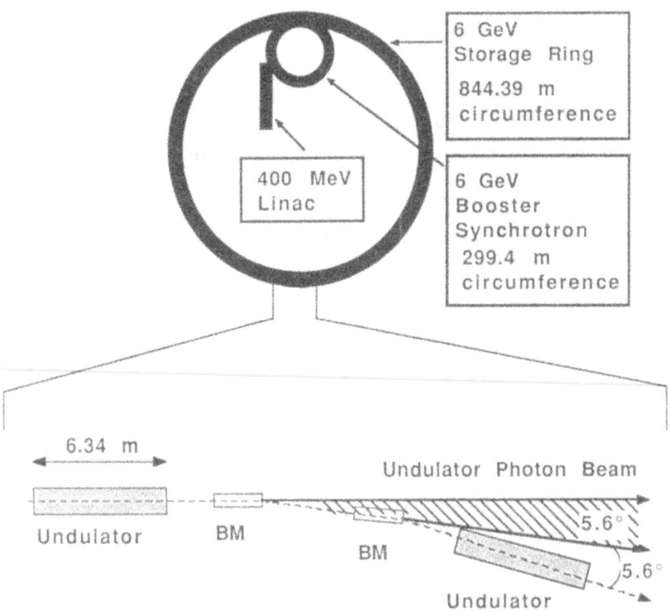

Fig. 4. Schematic design of the ESRF showing the synchrotron, the storage ring and the injector. The magnetic lattice has a 32 fold symmetry. Part of the lattice is shown in more detail below (without quadrupole and sextupole magnets). Photons can be observed along the axis of the undulator but also within an angle of 5.6° from the bending magnets.

2.2.3 Undulator Versus Wiggler

A decision between an undulator and a wiggler for SAXS experiments has to be based on the optimum flux at the sample for a given SAXS resolution and the necessity to tune the wavelength for anomalous dispersion studies. The beamprofile should furthermore be rather symmetric at the aperture slits in order to get a comparable SAXS-resolution both in the horizontal and vertical directions.

The photon emission of an undulator is concentrated into a central cone with half-width $x'(\sigma)$, $z'(\sigma)$. (observation direction $\theta = \varphi = 0$)

$$x' = (\sigma_x'^2 + \sigma_r'^2)^{1/2} \; ; \qquad z' = (\sigma_z'^2 + \sigma_r'^2)^{1/2} \tag{12}$$

where

$$\sigma_r' = (\lambda_n/L_i)^{1/2} \tag{13}$$

is the radiation cone emanating from a single electron and L_i the length of the undulator. The wavelength of the nth harmonics — λ_n — is given by (odd harmonics only; $\theta = \varphi = 0$): [13]

$$\lambda_n = (1/n)\,(\lambda_u/2\gamma^2)\,(1 + K^2/2) \tag{14}$$

The halfwidth of the source point is given by:

$$x = (\sigma_x^2 + \sigma_r^2)^{1/2} \; ; \qquad z = (\sigma_z^2 + \sigma_r^2)^{1/2} \tag{15}$$

where the single electron value $\sigma_r = 1/(4\pi)\,(\lambda_n L_i)^{1/2}$.

Corresponding values, together with the size of the beam at 30 m distance from the source point — calculated via Eq. 6 — are indicated in Table 2. ESRF

Table 2. Parameters of the electron beam, the photon beam at the source point (one std. dev.) and at 30 m distance from the source point (two std. dev.) for an undulator and a wiggler

		Undulator		Wiggler	
		β_x; β_z (m) 30; 11	β_x; β_z (m) 0.6; 2.5	β_x; β_z (m) 0.6; 2.5	
σ_x	(mm)	0.41	0.069	0.069	Electron
σ_z	(mm)	0.084	0.047	0.047	Beam
σ_x'	(mrad)	0.015	0.089	0.089	
σ_z'	(mrad)	0.007	0.013	0.013	
x	(mm)	0.41	0.069	0.084	Photon Beam
z	(mm)	0.084	0.047	0.067	at Source
x'	(mrad)	0.015	0.089	2.04	Point
z'	(mrad)	0.009	0.014	0.050	
S_1^x	(mm)	1.26	5.33	1.22	Photon Beam at
S_1^z	(mm)	0.56	0.85	3.0	Aperture Slits

specific undulators [19] — constructed according to the permanent magnet technology — consisting out of three 1.7 m long sections, with $\lambda_u = 3.4$ cm, a remanent field $B_r = 1.1$ [T] have been assumed. For $K \approx 0.7$ and a gap between the poles of ≈ 20 mm, the fundamental is at ≈ 8 KeV (≈ 1.5 Å).

The two undulators differ in the β-values which are defined by:

$$\beta_x = \sigma_x^2/\varepsilon_x\,; \qquad \beta_z = \sigma_z^2/\varepsilon_z \tag{16}$$

The β-values are fixed by the magnet lattice [13]. It can be seen that — for constant ε_x, ε_z — the high-β undulator is optimized for a small divergence, while the low-β is optimized for a small source point.

Evidently these type of sources result in a quite symmetric beamprofile at the aperture slits. The high β-undulator is, however, preferable if one wants to have the smallest possible aperture slits.

For a wiggler, the vertical source size is given by:

$$z = [\sigma_z^2 + (L_i^2/4)\,\sigma_r'^2]^{0.5} \tag{17}$$

where $\sigma_r' = 0.57/\gamma$ [20]. The vertical angular divergence is defined by Eq. 12.

In the horizontal direction, the electron beam excursion implies that $x' \approx 2K/\gamma$. The horizontal average source size is given as:

$$x = [((K \cdot \lambda_u)/(2\pi\gamma))^2 + (\sigma_x^2 + (L_i^2/4)\,\sigma_r'^2)]^{1/2} \tag{18}$$

An ESRF specific wiggler [19] with a total length of 2 m, a period of 10 cm, a K-value of 12 and a critical energy of 29 keV was assumed for the calculations.

Values corresponding to the size of the wiggler-beam at 30 m from the source point have been incorporated into Table 2. This type of source does not give a symmetric beam profile and has to be tailored by slits.

A comparison shows, that the area illuminated by a high-β undulator at $F_1 = 30$ m is ≈ factor 300 smaller than that of the wiggler discussed.

One can calculate approximately the flux of the wiggler, passing through a slit which has the size of the beamprofile of the undulator. Thus the flux distribution of a bending magnet — integrated over the vertical plane — is given by:

$$dFl/d\theta = 2.457 \cdot 10^{13} E \cdot I \cdot G_1(y) \tag{19}$$

where Fl is the flux, I the current [A] in the ring and the function $G_1(y)$ is given in literature [21].

The flux of the wiggler is obtained by multiplying Eq. 19 by the angle of the radiation in the horizontal plane and 2N. For $I = 0.1$ A one calculates a flux of $\approx 2 \cdot 10^{15}$ Ph/s/0.1% BW at 8 keV.

The total useful flux of an undulator is given by: [21]

$$Fl = 1.431 \cdot 10^{14} N \cdot Q_n \cdot I \tag{20}$$

Q_n is defined in Ref. 20. For the undulator specified above and 0.1 A electron current one calculates also a flux of $\approx 2 \cdot 10^{15}$ Ph/s/0.1% BW.

This implies that the flux emanating from a wiggler through a slit of the size of an undulator beam is $\approx$ factor 300 times smaller than the corresponding undulator flux. An undulator is therefore a better insertion device for intensity optimized experiments at small Q-values.

The possibility of wavelength tuning can be used to reduce the background scattering from heavy atoms. Anomalous dispersion studies furthermore demand a fine tuning of the wavelength with $\Delta E/E \approx 10^{-4}$. The wavelength of the fundamental and the harmonics of an undulator can be tuned via the variation of the gap of the magnets. Thus for the fundamental and an on-axis observation the variation of the wavelength is determined by K (Eq. 14), which depends on the gap via the magnetic field [21].

The tuning range of an ESRF undulator — to be used in the range 0.5 Å < λ < 2 Å — is shown in Fig. 5 [11, 22]. At E = 6 GeV this wavelength range can only be covered by a combination of the fundamental and third harmonics.

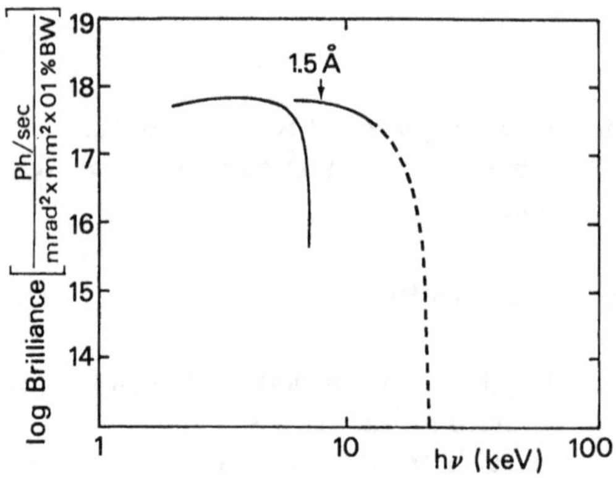

Fig. 5. Calculated variation of the brilliance of the undulator fundamental and third harmonics as a function of the energy upon tuning [22]

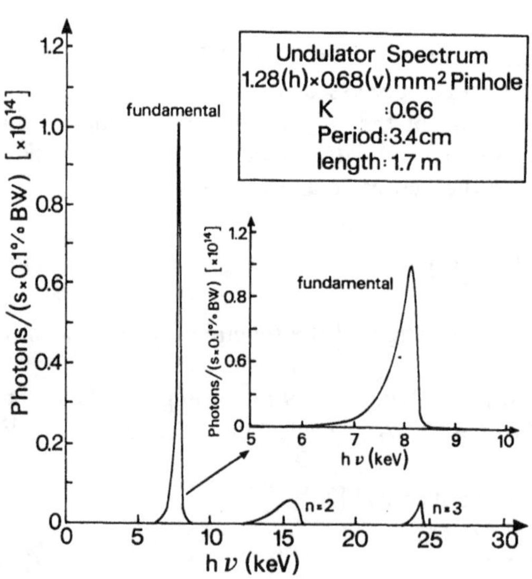

Fig. 6. Calculated undulator spectrum observed through a rectangular slit at 30 m from the source point. [23] The fundamental is shown in more detail in the inset

The relative bandwidth of the central peak of an undulator peak — observed on axis — is given approximately by:

$$\Delta E/E = 1/(N \cdot n) \tag{21}$$

where n is the order of the peak. For a 1.7 m undulator with $\lambda_u = 3.4$ cm, $\Delta E/E$ of the fundamental is therefore about $2 \cdot 10^{-2}$.

A calculated spectrum — observed through a rectangular slit of $1.28(h) \cdot 0.68(v)$ mm^2 — at a distance of 30 m and the undulator parameters $\beta_x = 30$ m, $\beta_z = 11$ m, $\lambda_u = 3.4$ cm, a gap of 24 mm, $K = 0.66$ and $\lambda_1 = 1.5$ Å shows the first three harmonics (Fig. 6) [23]. The inset shows the fundamental in more detail. The sawtooth shape is due to the 1D-periodicity of the magnetic structure. The value of $\Delta E/E \approx 5 \cdot 10^{-2}$ is larger than expected from Eq. 21 due to the angular divergence of the electron beam. Such calculated spectra agree quite closely to those observed at PEP [18] operating in an experimental high-brilliance mode.

3 Proposed Instrumentation

3.1 Introduction

Instrumentation presented below is based on a number of scientifically (1989) interesting areas. Although a principal agreement on a high flux, microfocus and anomalous dispersion camera has been reached, neither instrument has been defined as yet in detail.

The discussion will be restricted to the design goals, as related to present research interests. Questions related to heating of optical elements, response of detection systems or data aquisition systems will not be treated.

3.2 High Flux Camera

3.2.1 Design Goals

Applications of this camera will be in the area of time-resolved scattering on synthetic or biological supramolecular assemblies and in materials research (e.g. crystallization of glasses). The aims for the camera will be:

- small angle scattering up into the light scattering range together with wide-angle scattering capability
- flux increased by at least factor 10^2 as compared to cameras located at bending magnets
- wavelength range $0.5 < \lambda < 2$ Å

The limitations of an existing camera is shown in Fig. 7 for the case of a virgin poly[bis(*p*-methylphenoxy)phosphazene] (PBMPP) sample, which is heated through a T(1)-transition and cooled down again [24]. The experiment was performed on the Hasylab polymer beamline. Although the background has already been substracted, a peak at ≈600 Å can barely be resolved for the cooled down

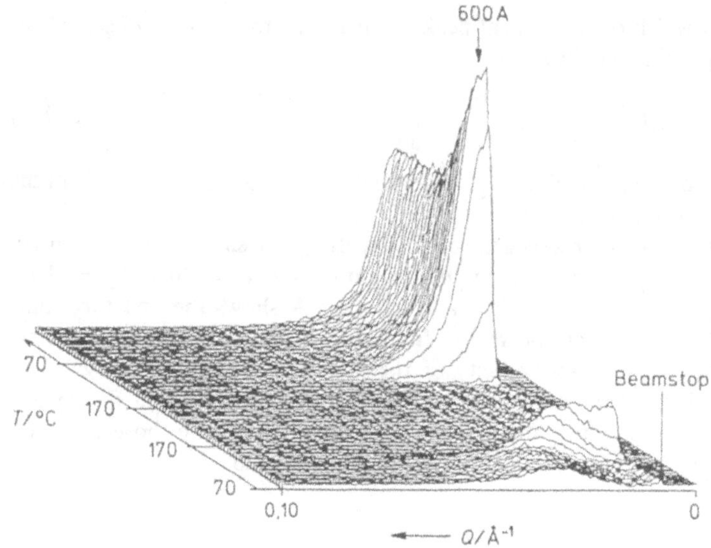

Fig. 7. Variation of SAXS patterns (8 s/frame) of virgin poly[bis(*p*-methylphenoxy)phosphazene] upon heating through T(1) ($\approx$ 130 °C) and cooling down [24]

sample. A better separation of such peaks from the area of stray radiation is therefore important.

There is no hope of getting direct structural information on larger assemblies — like spherulites (μm-range; Fig. 8) — which are formed in the course of crystallization processes in polymers with this camera. Such an information is, however, useful in order to understand macroscopic properties [25].

Spherulitic crystallization is studied at present with light scattering methods [26]. An advantage of X-ray scattering — as compared to light scattering — is that turbid or opaque samples can be studied and that the accessible Q-range extends up to the wide-angle scattering range. Complimentary wide-angle and small-angle scattering

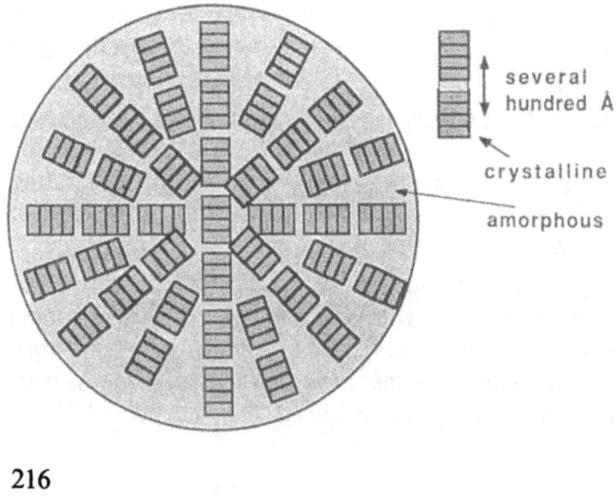

Fig. 8. Schematic design of a spherulite which is formed in the course of polymer crystallization. Spherulites contain blocks of crystalline zones, embedded in an amorphous matrix. The size can be in the μm range and larger

information have been shown to be necessary in order to understand crystallization processes [27, 28].

In molecular biology, certain structural features may not show up in electron microscopy. This is, for example, the case for the cycloskeletal lattice of frog sartorius muscle [29].

Although the light scattering range is also accessible by the Bonse-Hart camera [2], this type of optics is not well suited for real time experiments.

It is difficult to define for the mirror/monochromator instrument a lower limit in Q-range. One possibility would be to demand that the beam cross section at the sample should not exceed ≈ 0.25 mm^2 [2], which is important for real time solution scattering and corresponds to ≈ 0.25 µl for 1 mm path length. This implies $F_2 \leqq 20$ m ($L_{max} \approx 3$ µm; see below). One would need, however, even larger F_2-values if one would like to study the critical behaviour of PVME/PSD polymer-blends in real time. Thus neutron scattering studies on static samples had to be extended down to scattering vectors of 10^{-5} Å^{-1} in order to see an interference peak, assumed to be due to an agglomeration of domains [30] (Fig. 9). If one could use larger sample areas (e.g. polymer foils [4]) then an increase in F_2 poses no problems.

A high flux at the sample is important for the study of conformational transformations in stopped flow experiments on biological solutions. Thus molecular association can presently be easily detected on a timescale down to the msec range. Modelling of such processes has been possible for example for the case of microtubuli formation [31]. In other cases, where less association occurs, only the central scattering could be evaluated which resulted only in kinetic information [8]. Mechanistic and structural studies based on a larger Q-range will therefore demand a higher flux.

There is also interest in studying irreversible structural processes on a shorter time scale than presently accessible (depending on the Q-range covered between ms to s-range).

Finally molecular weight fractions of polymers or biological materials which are difficult to prepare in large quantities could be studied more readily at higher fluxes than presently available.

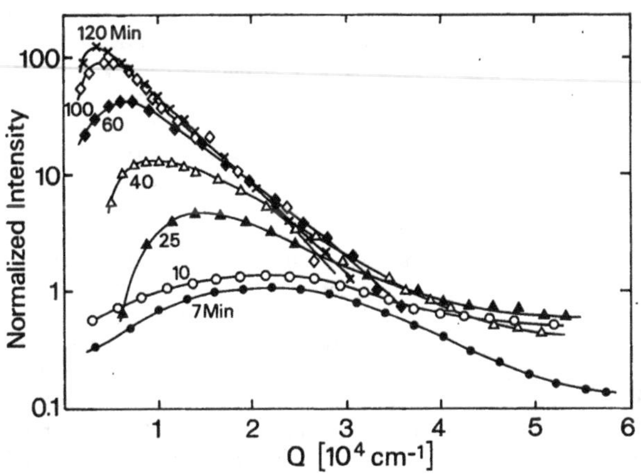

Fig. 9. Interference peak due to spinodal decomposition of PSD/PVME blends at different annealing times [29]

3.2.2 Design Parameters

A camera with a bent mirror/bent monochromator optics (Fig. 10a) has the advantage of a high flexibility in its focusing conditions and allows to minimize the beam cross section at the sample for a given SAXS-resolution, which is especially of interest for small samples like single muscle fibres. This implies that a high-β undulator will be used (Table 2). Note that beam compression via an asymmetric cut monochromator [32] is not necessary in view of the highly symmetric source.

The possibility of wavelength tuning ($0.5 < \lambda < 2$ Å) is of interest in order to suppress fluorescence scattering of heavy atoms, i.e. during the crystallization of inorganic glasses. Due to geometric constraints this will be possible for this type of camera only up to $F_2 \approx 8$ m. Longer F_2-values can only be realized for a fixed wavelength of ≈ 1.5 Å [11].

The limit of diffuse scattering (L) — which corresponds to the SAXS-resolution — has been calculated for this type of optics for various camera length (Fig. 11a). A comparison of the optimum in L (L_{max}) with that of the polymer beamline at HASYLAB is shown in Fig. 11b. At a comparable $F_2 \approx 9$ m one obtains an increase in L_{max} by $\approx$ factor 15. The size of the beam at the sample is shown in Fig. 11c.

If one restricts the length of the camera to be within the experimental hall ($F_2 \approx 25$ m) then $L_{max} \approx 3 \cdot 10^4$ ($Q_{min} \approx 2 \cdot 10^{-4}$ Å^{-1}). There is no reason, however, why the camera could not be extended beyond the experimental hall in order to reach Q_{min} values of $\approx 10^{-5}$ Å^{-1} and less in case the samples could be sufficiently large or one would be prepared to sacrifice intensity by reducing the beam size. Note that a proposed topography beamline has a length of ≈ 400 m [11].

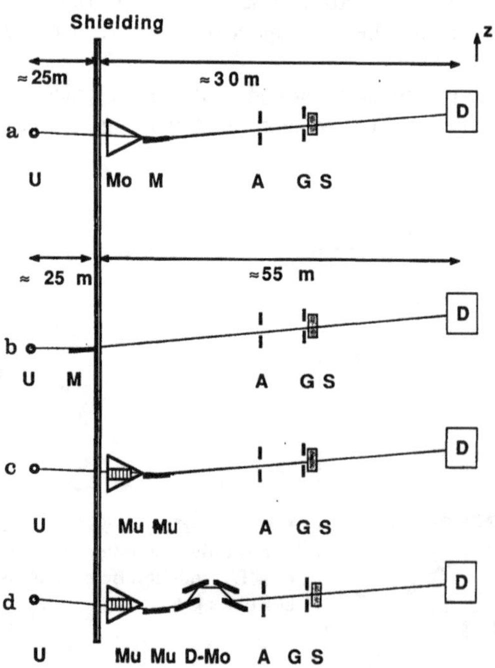

Fig. 10a–d. Different types of SAXS-cameras discussed in text.
a. Pinhole camera with bent mirror and bent monochromator optics. **b.** Schematic design of pinhole camera with deflection into the vertical plane. **c.** Pinhole camera with orthogonal bent multilayer optics. **d.** Pinhole camera with orthogonal bent multilayer optics and two double monochromators
Abbr.: U: undulator; M: mirror; Mo: monochromator; D-Mo: double monochromator; Mu: multilayers; A: aperture slits; G: guard slits; S: sample; D: detector. The length scale is determined by the insertion of the camera into the experimental hall. The different length of the mirror/monochromator camera is due to the horizontal deflection by $2\theta \approx 27°$ ($\lambda \approx 1.5$ Å)

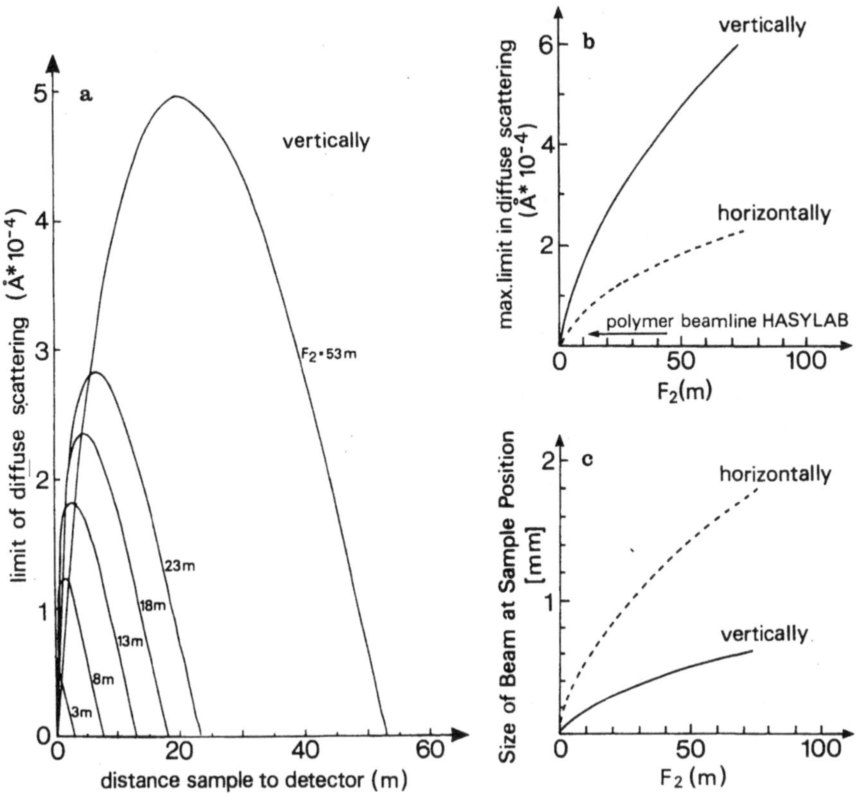

Fig. 11 a. Calculation of the limit of diffuse scattering in the detector plane. The wavelength is 1.5 Å. **b.** Optimum in L-values as a function of F_2. **c.** Illuminated area at the sample as a function of F_2 for optimum L-values

In order to cover both the wide-angle and small-angle scattering range, at least two detectors at different distances to the sample will have to be synchronized.

The photon flux at the sample can be calculated from:

$$Fl_s = Fl \cdot ([\Delta E/E]/10^{-3}) \cdot W \cdot P \cdot A \tag{22}$$

where Fl is given in Eq. 20, W is the absorption by windows (≈ 0.5), P the polarisation loss (≈ 0.9) and A the loss by absorption (≈ 0.9) A flux of $\approx 8 \cdot 10^{13}$ Ph/s is thus calculated which is $\approx$ factor 800 larger than that of a bending magnet SAXS-beamline [32].

The flux is limited for this type of optics by the energy resolution of the monochromator (e.g. $\Delta E/E \approx 3 \cdot 10^{-4}$ for Ge(111)). Such an energy resolution is, however, not necessary for most SAXS-experiments. Thus the width — w — of a reflection in the diffraction plane may be written at small angles as: [1]

$$w = (a^2 + \lambda^2 c^2 L_2^2 + [\Delta E/E\theta L_2]^2)^{0.5} \tag{23}$$

c corresponds to the intrinsic width of the reflection and θ to the Bragg angle.

The reflection broadening — $\Delta\theta_{tot}$ — may be written as:

$$\Delta\theta_{tot} = (\Delta\theta_p^2 + \Delta\theta_d^2 + \Delta\theta_m^2)^{0.5} \quad (24)$$

where $\Delta\theta_p^2$ is due to the particle size, $\Delta\theta_d^2$ due to disorder and $\Delta\theta_m^2$ due to the mosaicity.

For solid or dissolved synthetic polymers (with well defined molecular weight fractions!) or biopolymers, the last term can be neglected, while the first two terms correspond to an apparent particle size.

$$c^{-1} \approx \lambda[(\Delta\theta_p^2 + \Delta\theta_d^2)]^{-0.5} \quad (25)$$

For the source parameters of the ESRF and ideal focusing conditions, the bandwidth could be increased from $\approx 3 \cdot 10^{-4}$ by $\approx$ factor 100 without an influence on the peakwidth up to $c^{-1} \approx 2 \cdot 10^4$ Å. This is shown in Fig. 12 for a 10^3 Å and a 10^4 Å peak with the instrumental parameters as indicated in the figure. Deviations from an ideal optics which blur the focus will act like an increase of $\Delta E/E > 3 \cdot 10^{-2}$. This is shown in Fig. 12 for an increase of a_z to 1 mm (10^4 Å peak).

There are two options to increase the flux at the sample for a double focusing optics:

(i) the bandwidth could be increased to $\approx 10^{-3}$ via an imperfect monochromator crystal but with an associated increase of diffuse scattering
(ii) the full width of of an undulator peak is used (Fig. 6)

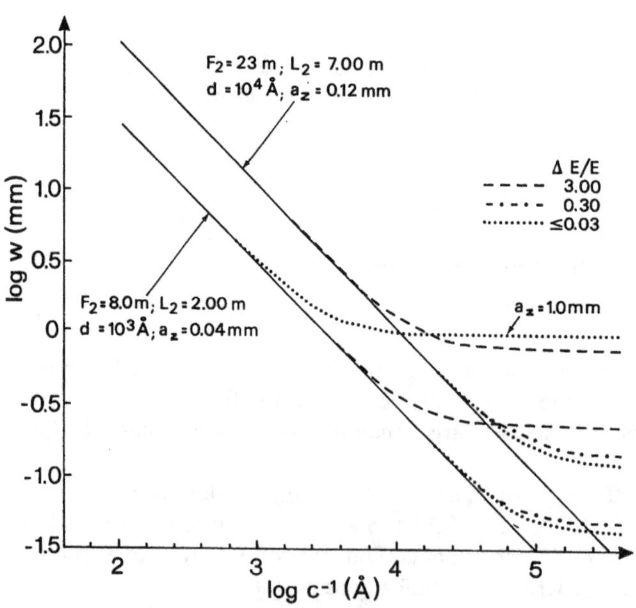

Fig. 12. Variation of log w (w: width of d = 10^3 Å and d = 10^4 Å reflections in the detector plane) as a function of log c^{-1} (c^{-1}: apparent particle size) for different bandwidth $\Delta E/E$. The instrumental parameters are shown in the figure

The simplest optics making use of the full undulator peak is a pinhole camera without focusing elements. The only optical element necessary is a plane mirror in order to elevate the beam out of the plane of the storage ring, to reduce the background due to Bremsstrahlung and to cut the harmonics (Fig. 10b).

This type of camera is, however, incovenient for scattering on very small samples — like muscle fibers — which demand an optimization of the flux and a minimalization of the beam cross section at the sample.

Of more general use might be the use of multilayers — e.g. W/C — in order to tailor the SR-spectrum. [33, 34, 35]. The bandwidth acceptance is given by the number of layers — N — via:

$$\Delta E/E = 1/N \tag{26}$$

and can be matched to the bandwidth of an undulator peak. The peak reflectivity is at 1.5 Å ≈ 60%. This suggests an optics with bent multilayers for horizontal and vertical focusing instead of a monochromator/mirror optics (Fig. 10c).

As the angle of reflection of such multilayers is only a few degrees at $\lambda = 1.5$ Å, the camera would be nearly linear, which is an advantage for the insertion into the experimental hall and would allow to reach already $L_{max} \approx 1 \cdot 10^{-4}$ Å^{-1}. The footprint on the multilayers would be at 1.5 Å about factor 10 larger than for a Ge(111) crystal, which would allow a better distribution of the heatload.

The flux at the sample is calculated to be $\approx 8 \cdot 10^{15}$ Ph/s (Eq. 22). This increased flux could find two applications.

(i) an improved statistics for realtime scattering studies on weakly scattering samples, e.g. onset of decomposition processes or more rapid experiments in case appropriate detection systems become available. Thus the present timescale of SAXS experiments on solid polymers is generally a few sec./spectrum [4]. Here one would expect measuring times per spectrum in the 50–100 μs range.

(ii) a reduction of the aperture slits and thus a further decrease of the range of diffuse scattering. (Eq. 2)

For experiments demanding a higher ΔE/E resolution (possibly for experiments on highly ordered superlattices), a double monochromator could be added. The use of two double monochromators (Fig. 10d) would allow to maintain the direction of the beam.

Prior to the realization of such an optics, more information on the use of multilayers as optical element in SAXS (e.g. background, focusing properties and thermal stability) has, however, to be obtained. In view of the limited stability of some (especially biological) samples [36], such a camera must be equipped with an absorber in order to align samples at a reduced flux and fast shutters.

3.3 Microfocus Camera

3.3.1 Design Goals

Applications of a microfocus camera would be in areas such as:

- SAXS/WAXS studies of synthetic or biological fibrils in the μm-range and less, or complex morphological structures encountered in processed polymer foils.

— Scanning microfocus SAXS/WAXS probe
— Diffraction on small crystals with large unit cells (e.g. virus crystals or proteins) or small polymer crystals (e.g. Polyethylene).

Thus morphological structures like spherulites (Fig. 8), which appear in the course of polymer crystallization, could possibly be studied individually.

3.3.2 Design Parameters

The instrument aims at the smallest possible focus, which demands an optimized source brilliance. This implies again the use of an undulator as an insertion device and a demagnifying optics. In this case a low-β undulator (Table 2) would be advantageous.

A double mirror optics with fixed focal length appears at present to be best suited in order to minimize the focus.

Fig. 13 shows a possible beamline at the ESRF with a flat gold covered pre-mirror and a focusing toroidal mirror [11]. The flat mirror is necessary in order to lift the beam out of the plane of the storage ring where a high background due to Bremsstrahlung is expected. It furthermore cuts the higher harmonics of the undulator.

The demagnification attainable from mirror optics is limited by surface roughness — which creates diffuse scattering [37] and can be limited by slits — and by surface slope error (∂') — which blurs the focus.

An effective size of the focus can be defined as:

$$a^x_{eff} = 2[(x/2)^2 + (2 \cdot F_2 \cdot \partial')^2]^2 \; ; \quad a^z_{eff} = 2[(z/2)^2 + (2 \cdot F_2 \cdot \partial')^2]^2 \tag{27}$$

For a demagnification ratio $F_2/F_1 = 0.34$ a focal spot size of $20 \cdot 14\ \mu m^2$ (σ^2) has been calculated [11]. The limit of diffuse scattering would in this case be at $Q \approx 6 \cdot 10^{-4}$ Å^{-1} ($L_{max} \approx 1\ \mu m$).

In order to avoid blurring, such a demagnification would demand a mirror with a surface slope error of ≈0.1″ which is at the limit of the technological possibilities.

Alternatively one could use multilayers — deposited on mirrors — for focusing and tailoring of the spectrum. This would allow to decrease the distance of the two optical elements. A high demagnification and a fixed focal length would be necessary.

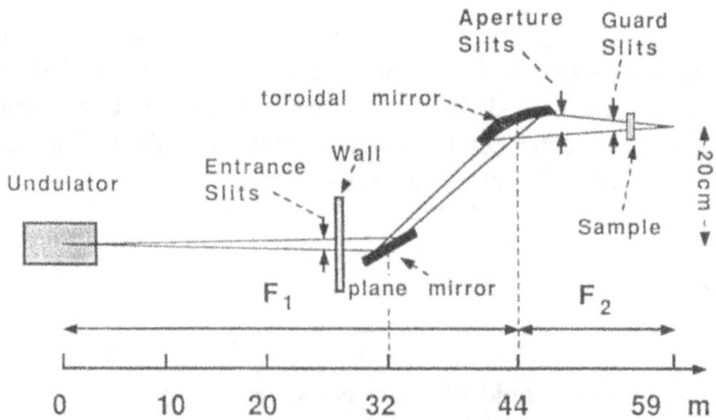

Fig. 13. Double mirror optics SAXS-camera [11]

In SR-scattering, a focus of $\approx 10 \cdot 40\,\mu m^2$ has been realized for a microprobe instrument, using two cylindrical mirrors (Kirkpatrick-Baez geometry [38]) as substrates. Such an optic appears to be suitable to reach focal spot sizes of $\approx 1 \cdot 1\,\mu m^2$ [39].

If one could reach spotsizes at the sample of $\approx 0.1 \cdot 0.1\,\mu m^2$ then one would get into the range of the electron density fluctuation due to the morphology of lamellar polymers. The beamsize would be much smaller than larger morphological units — like spherulites — which contribute to the scattering at lower angles. It should therefore be possible to reduce the background scattering at smaller angles by selecting the appropriate demagnification of the focus.

Questions of sample stability and sample alignment will be still more dominating than for the realtime camera. Organic samples would be less affected than inorganic samples due to the difference in absorption coefficient μ. Thus the temperature of the sample may be calculated from: [11]

$$P_a = \varepsilon\sigma(T_2^4 - T_1^4) \tag{28}$$

where P_a is the power absorbed per unit area, ε the emissivity, σ the Stefan-Boltzmann constant, T_1 the starting and T_2 the final sample temperature. P_a may be calculated from:

$$P_t = P_0 e^{-\mu t} \tag{29}$$

where

$$P_a = P_0 - P_t \tag{30}$$

Calculations of the temperature reached as a function of the sample thickness — t — and the parameters $P_0 = 30\,W/mm^2$, $\varepsilon = 0.1$, $\sigma = 5.67 \cdot 10^{-14}\,Watt/mm^2$, $T_1 = 300\,K$ are shown in Fig. 14 for Carbon ($\mu = 4\,cm^{-1}$) and Silicon ($\mu = 144\,cm^{-1}$). While Si is heated to ≈ 1000 K already for a few hundred Å thickness, C reaches only a few

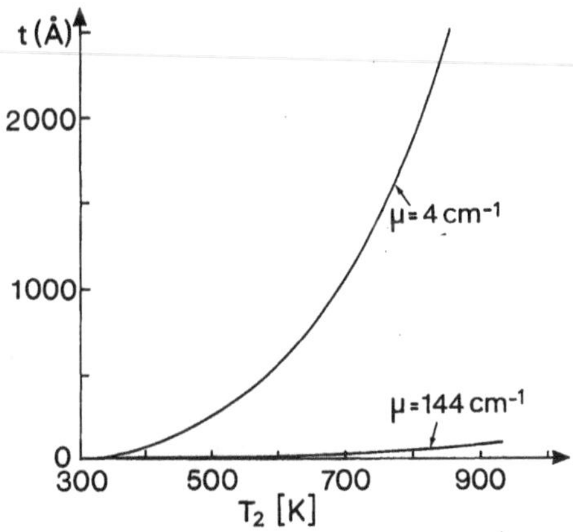

Fig. 14. Calculated variation of the temperature (T_2) of carbon and silicon as a function of the sample thickness (t) for a power load of 30 W/mm²

hundred K. Both values are, however, too high for actual experiments and suggest the necessity of short time exposures (fast shutters!) and possibly cryotechniques for sample handling.

3.4 Anomalous Dispersion Camera

3.4.1 Design Goals

Anomalous dispersion experiments allow to change the atomic form factor — f — via the wavelength dependance of the dispersion corrections f′ and f″.

$$f = f_0 + f' + if'' \tag{31}$$

where f_0 is the formfactor term for photon energies higher than the binding energy of the innermost shell. Significant variations of f-values are observed at the specific atomic absorption edges, K, L, M ... within a range of $\Delta E/E \approx 10^{-2}$ (Fig. 15). The variation of the contrast of specific atoms relative to its environment can be used for structural studies.

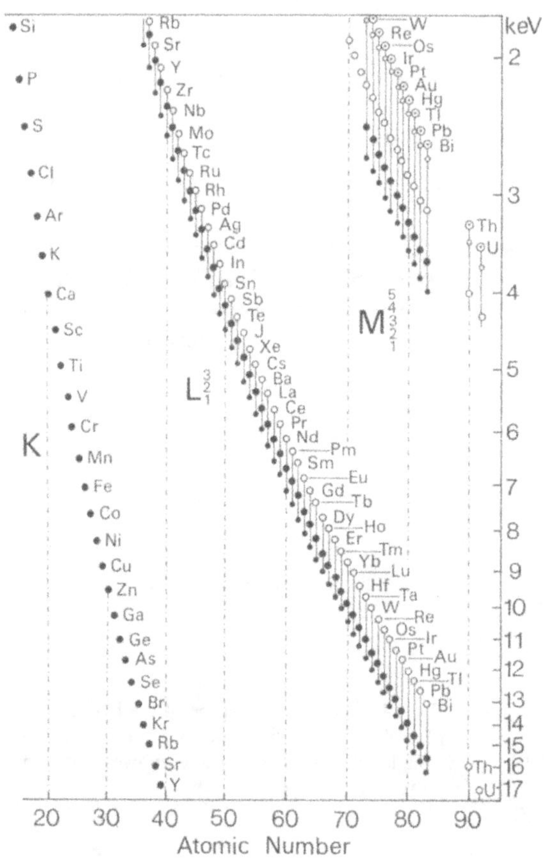

Fig. 15. Variation of the energy of the K,L,M-edges of elements with the atomic number. The structure of the edges is indicated [40]

Anomalous dispersion small-angle X-ray scattering (ASAXS) experiments have been used to study alloys and polymers and biopolymers [9, 10, 41]. Thus from the determination of partial structure factors information on the repartition of elements during demixing of alloys was obtained. Slow kinetic studies have also been reported [9]. Fig. 16 shows the scaled partial structure factors P_{Me-Me} (Me = Fe, Ni, Cu) of an Cu—Ni—Fe alloy for three aging times. A kinetic law with $Q_1 = t^{-a}$ was assumed. The scaled structure function is defined by: [42]

$$P_{Me-Me}(x, t) = Q_1^3(t)\, S_{Me-Me}(Q, t)/\Sigma S_{Me-Me}(Q, t)\, Q^2\, dQ \tag{32}$$

with $x = Q/Q_1(t)$. S_{Me-Me} is the partial structure factor. The presently available flux limits such experiments to few data points at late stages of unmixing.

An ASAXS-camera should make possible:

- study of K-absorption edges for elements up to $Z \approx 50$
- factor $\approx 10^2$ higher flux than presently available and hence the possibility of ASAXS studies of early stages of phase separation and crystallization or very diluted (e.g. biological) systems with heavy atoms
- medium SAXS-resolution up to $L_{max} \approx 4000$ Å ($Q_{min} \approx 10^{-3}$ Å^{-1})

3.4.2 Design Parameters

The classical camera for anomalous dispersion studies is the double crystal diffractometer, using perfect Si or Ge-crystals [43, 44] (Fig. 17a). The energy resolution of

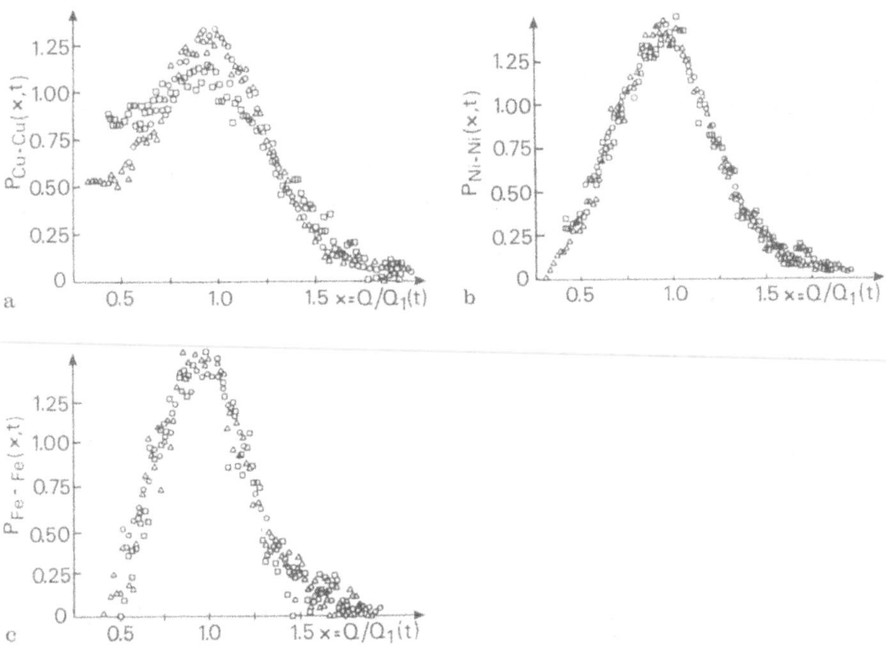

Fig. 16a–c. Scaled partial structure factors P_{Fe-Fe}, P_{Ni-Ni}, P_{Cu-Cu} for a partially demixed Cu—Ni—Fe alloy at three aging times: 3 h (*squares*), 9 h (*triangles*), 56 h (*circles*) [9]

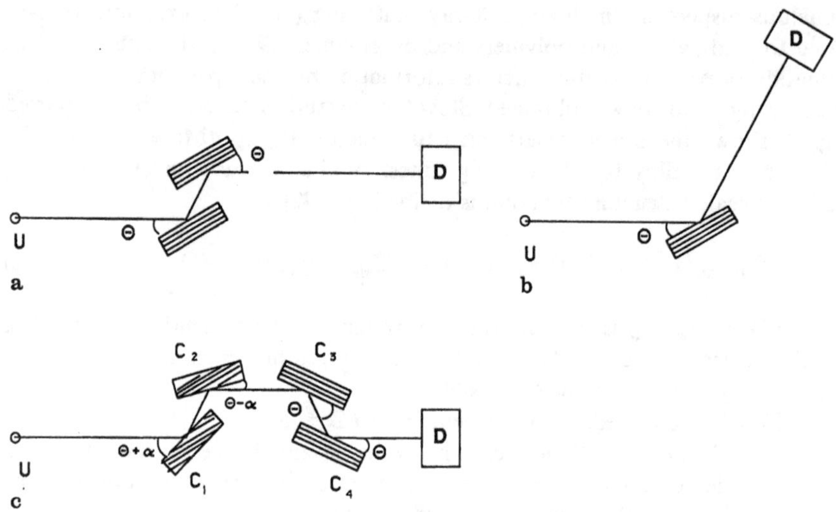

Fig. 17a. Double monochromator camera for SAXS studies. The orientation of the reflecting planes is schematically indicated. **b.** Single monochromator camera for SAXS studies. **c.** Four-crystal monochromator setup

$\approx 10^{-4}$ obtained by such a camera allows to tune the wavelength with sufficient precision.

The cutoff by Be-windows at ≈ 3 Å permits to study elements down to $Z \approx 20$ (Ca; K-edge). Elements like Cl, S, P — which would be of interest to study in synthetic and biological polymers — would demand wavelength up to ≈ 6 Å (≈ 2 keV) and hence a windowless camera. The ESRF has at this energy a still higher spectral brilliance from an undulator than the proposed 2 GeV Trieste machine [17, 45].

The peak reflectivity of a Si-crystal is at 6 Å only ≈ 0.15, which reduces the reflectivity to $2 \cdot 10^{-2}$ in a double crystal diffractometer. Consequently, a recently developed diffractometer for softer wavelength employs a single crystal optics [46] (Fig. 17b). As experimental experience with softer X-rays in SAXS is still very limited, the following discussion will be restricted to the double crystal diffractometer camera with $\lambda_{max} \approx 3$ Å.

In order to cover the K-edges up to $Z \approx 50$, the energy range must extend up to ≈ 30 keV. If one wishes to keep the advantages of an undulator for a pinhole type camera, harmonics up to $n = 7$ have to be used (Fig. 18). This implies that — at a fixed undulator gap — an undulator peak with $\Delta E/E \approx 3 \cdot 10^{-2}$ is scanned by the double monochromator which has a bandwidth of $\Delta E/E \approx 2 \cdot 10^{-4}$. Larger energy ranges are accessible via several undulator settings followed each time by a double monochromator scan.

At present only a pinhole geometry without focusing elements has been studied. The SAXS resolution would in this case be more limited as for the real-time camera with $L_{max} \approx 4500$ Å. Focuring in the horizontal plane would, however, be possible by sagittal focusing via the second crystal [47].

A flux at the sample of $\approx 8 \cdot 10^{12}$ Ph/s is calculated at 8 keV acc. to Eq. 22.

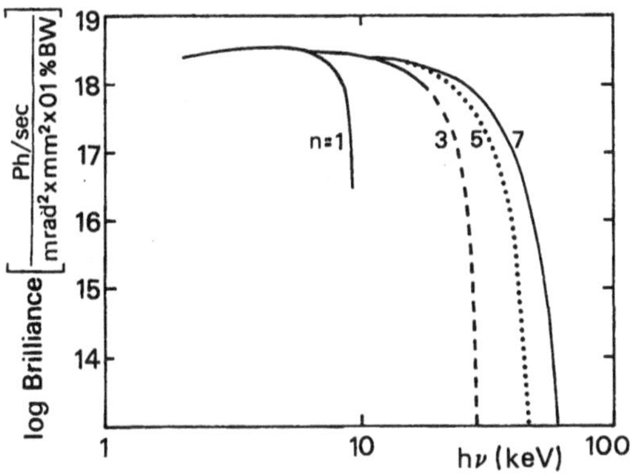

Fig. 18. Variation of brilliance as a function of energy for undulator harmonics up to n = 7 [22]

The choice in optics is also in this case not finalized. Thus one might consider to use a double monochromator in an arrangement shown in Fig. 17c. [48] Crystals C_1 and C_2 would be Si(111) crystals cut asymmetrically under an angle $+\alpha(C_1)$ and $-\alpha(C_2)$ while crystals C_3, C_4 would have a symmetric cut.

An asymmetry factor — A — can be defined according to: [49]

$$A = (\sin \theta_B - \alpha)/\sin (\theta_B + \alpha) \tag{33}$$

where θ_B is the monochromator Bragg angle. The width of the beam after an asymmetrically cut crystal — w_a — is related to that before this crystal — w_b — by:

$$w_a = w_b/A \tag{34}$$

This would result in an expansion of the beam at C_1 and a compression at C_2 and hence a better distribution of the heatlocad. Assuming for example $\alpha = 11°$ at 8 keV, the footprint could be increased at C_1 by factor ≈ 10.

The bandwidth obtained from such a double crystal arrangement is: [49]

$$\Delta E/E = (\Delta\theta + [\omega_s/A^{0.5}]) \cot \theta_B \tag{35}$$

where $\Delta\theta$ is the angular acceptance and ω_s the intrinsic width of the crystal. As $\Delta E/E$ will be increased — as compared to a double crystal monochromator without asymmetric cut — a further double crystal monochromator is used in order to obtain $\Delta E/E \approx 2 \cdot 10^{-4}$ (Si-111).

3.5 Other Cameras

The examples for SAXS-beamlines indicated above are far from exhaustive and brief mention should be made of other potential SAXS techniques which are already in use or under development:

— The Bonse-Hart camera which has already been mentioned above is capable of reaching ultrasmall Q-values but is much more compact as compared to the realtime camera described above [2, 50]. It cannot, however, be used for fast realtime experiments.

— High-energy small-angle scattering at photon energies of 50 keV and more could be used to study heavily absorbing materials and in particular samples in a heavily absorbing environment, such as a pressure or a flow cells [51].
— Grazing-incidence SAXS could be used to study thin polymeric films at surfaces. Organic surface layers — like Langmuir — Blodgett films — are being studied at present only by WAXS techniques [52].

4 Conclusions

The ESRF will allow a considerable progress in small-angle scattering instrumentation in terms of accessible Q-range, size of focus and flux at the sample. New domains of SAXS experiments can be expected to become important in the future as experimental developments proceed.

Systematic studies on the optics, detection systems and the behaviour of materials under high photon flux have to be undertaken prior to the realization of future instrumentation at the ESRF. This will have to be done in close contact with national facilities and future users.

Acknowlegdement. Helpful discussions with J. Bordas, R. Coisson, P. Elleaume, A. Freund, J. Hastings, A. Luccio, A. Miller, B. Stephenson, C. Williams and H. G. Zachmann are acknowledged.

5 References

1. Rosenbaum G, Holmes KC (1980) In: Winick H, Doniach S (eds) Synchrotron radiation research, Plenum, New York
2. Koch MHJ, Bordas J (1983) Nucl. Instr. Meth. 208: 461
3. Koch MHJ (1988) Makrom Chem, Makrom Symp, vol 15: 79
4. Elsner G, Riekel C, Zachmann HG (1985) In: Kausch J (ed) Advances in polymer science, vol 67, Springer, Berlin Heidelberg New York
5. Wignall GC, Crist B, Russell TP, Thomas EL (eds) (1987) Scattering, deformation and fracture in polymers, Materials Research Society Symposia Proceedings, vol 79, MRS, Pittsburgh PA
6. Stephenson GB (1982) SSRL Report 82/05
7. Huxley HE, Faruqi AR, Bordas J, Koch MHJ, Milch JR (1980) Nature, 284: 140
8. Inoko Y, Kihara H, Koch MHJ (1983) Biophysical Chem. 17: 171
9. Lyon O, Simon JP (1987) Phys. Rev. B35: 5164
10. Derian PG, Williams C (1987) Proceedings of the Yamada Conference
11. Foundation Report (The Red Book) (1987) European Synchrotron Radiation Facility, B.P. 220, F-38043 Grenoble Cedex
12. Glatter O, Kratky O (eds) (1982) Small Angle X-ray Scattering, Academic Press, London
13. Krinski S, Perlman ML, Watson RE (1983) In: Koch EE (ed) Handbook on synchrotron radiation, vol 1A, North Holland, Amsterdam
14. Wiedemann H (1986) Nucl. Instr. and Meth. in Phys. Res. A246: 4
15. Brefeld W (personal communication)
16. Walker R (personal communication)
17. Scientific and technological applications of synchrotron radiation, Proceedings of the Workshop in Miramare-Trieste, 1987

18. Workshop on PEP as a Synchrotron Radiation Source, October 20–21, 1987
19. Elleaume P (1987) Design considerations for the insertion devices and beam front ends of the ESRF, Madison Conference on Synchrotron Radiation, extended abstracts, ESRF-Report, CONF-87-08
20. Coisson R (1985) Proceedings of SPIE, 582: 185
21. Kwang-Je Kim (1986) In: X-Ray Data Booklet, Lawrence Berkeley Laboratory
22. Computer code by R. Coisson (University of Parma)
23. Computer code by P. Elleaume (ESRF)
24. Magill JH, Riekel C (1986) Makrom. Chem. Rapid Comm. 7: 287
25. Baer E, Hiltner A, Keith HD (1987) Science 235: 1015
26. Stein RS, Wilkes GL (1975) In: Structure and properties of oriented polymers, Applied Science Publishers, London
27. Zachmann HG, Wiswe D, Gehrke R, Riekel C, (1985) Makrom. Chem. Suppl. 12: 175
28. Saijo K, Hashimoto T (1987) In: Wignall CD, Crist B, Russel TP, Thomas EL (eds) Mat. Res. Soc. Symposia Proc., vol 79, MRS, Pittsburgh Pa
29. Nave C, Diakun GP, Bordas J (1986) Nucl. Instr. and Meth. in Physics, A 246: 609
30. Schwahn D, Mortensen K, Yee-Madeira H (1987) In: Wignall CD, Crist B, Russel TP, Thomas EL (eds) Mat. Res. Soc. Symposia Proc., vol 79, MRS, Pittsburgh PA
31. Mandelkow E, Mandelkow EM, Bordas J (1983) J. Mol. Biol. 167: 179
32. Hendrix J, Koch M, Bordas MHJ (1979) J. Appl. Cryst. 12: 467
33. Bilderback DH, Larson BM, Barbee TW, Ice GE, Sparks CJ (1983) Nucl. Instr. Meth. 208: 251
34. Thompson AC, Wu Y, Underwood JH, Barbee TW (1987) Nucl. Instr. and Meth. in Physics, A255: 603
35. Stephenson GB (1988) Nucl. Instr. and Meth. in Physics, A 266: 447
36. Bordas J, Mandelkow EM (1983) In: Sha'afi RT, Fernandez SM (eds) Fast methods in physical biochemistry and cell biology, Elsevier/North Holland Bioch., Press, p 137
37. Matsushita T, Ishikawa T, Kohra K (1984) J. Appl. Cryst., 17: 257
38. Kirkpatrick P, Baez AK (1948) J. Opt. Soc. Am., 38: 766
39. Thompson AC, Wu Y, Underwood JH, Barbee TW (1987) Nucl. Instr. and Meth. in Physics, A255: 603
40. Stuhrmann HB (personal communication)
41. Stuhrmann HB, Notbohm N (1981) Proc. Natl. Acad. Sci. USA 78: 6216
42. Binder K, Heermann D (1985) Scaling phenomena in disordered systems, Plenum, New York
43. Dubuisson JM, Dauvergne JM, Depautex C, Vachette P, Williams CE (1986) Nucl. Instr. Meth. A246: 636
44. Haubold HG, Jaeger W, Lengeler B: ESRP-IRI-8/83
45. Elleaume P: ESRF-SR/ID-87-14
46. Stuhrmann HB (1982) Makrom. Chem. 183: 2501
47. Mills DM, Henderson C, Battermann BW (1986) Nucl. Instr. Meth., A246: 377
48. 7 GeV Advanced Photon Source Conceptual Design Report, Argonne National Laboratory, ANL-87-15 (1987)
49. Matsushita T, Hashizume H (1983) In: Koch EE (ed) Handbook on synchrotron radiation, vol 1A, North-Holland, Amsterdam
50. Bonse U, Pahl R, Nusshardt R (1987) Acta Cryst. A43: C-259
51. Riekel C (1988) In: Freund AK (ed) Workshop on applications of high energy X-ray scattering, ESRF, May 1988
52. see for example: Kjaer K, Als-Nielsen J, Helm CA, Laxhuber LA, Möhwald H (1987) Phys. Rev. Lett., Vol. 58: 2224

Author Index Volume 151

Author Index Vols. 26–50 see Vol. 50
Author Index Vols. 50–100 see Vol. 100
Author Index Vols. 101–150 see Vol. 150

The volume numbers are printed in italics

ISBN 3-05-500769-7

www.ingramcontent.com/pod-product-compliance
Lightning Source LLC
LaVergne TN
LVHW060624300926
849162LV00008B/1705